DEVELOPING AREAS RESEARCH GROUP
THE ROYAL GEOGRAPHICAL SOCIETY
(WITH THE INSTITUTE OF BRITISH GEOGRAPHERS)

DARG Regional Development Series No. 1

Development as Theory and Practice

Current perspectives on development and development co-operation

edited by David Simon and Anders Närman

 LONGMAN

Addison Wesley Longman Limited
Edinburgh Gate
Harlow
Essex CM20 2JE
United Kingdom
and Associated Companies throughout the world.

Visit Addison Wesley Longman on the world wide web at:
http://www.awl-he.com

First published 1999

ISBN 0 582 41417 2

British Library Cataloguing-in-Publication Data
A catalogue record for this book is available from the British Library

Library of Congress Cataloging-in-Publication Data
A catalogue record for this book is available from the Library of Congress

Set by 7 in 10/11 Palatino
Printed in Malaysia

Contents

Contents

Figures

Tables

Contributors

Françoise Barten holds an MSc in Public Health, London and a PhD Medical Sciences, Nijmegen. Her main research interest is in urban health systems and occupational health in small-scale industries. She is a lecturer at the Nijmegen Institute of International Health, and a consultant for WHO and various Dutch development agencies.

María Verónica Bastías is the Representative of Norwegian People's Aid (NPA) in South America, based in Santiago de Chile. Before this she was a project manager for Fundacion Nuestros Jovenas in Ecuador and a consultant for international agencies in Chile. She is an architect, specialising in regional planning, local development and co-operation. Further to this, she has taken specialised courses at universities in Chile, Ecuador, Colombia, Argentina and Norway.

Reginald Cline-Cole teaches geography at the Centre of West African Studies at the University of Birmingham, UK. He has published several studies on rural energy, forestry and land in West Africa, including the jointly-authored *Wood fuel in Kano*, and is currently co-editing *Contesting Forestry in West Africa*.

Ali de Jong is researcher and lecturer at the Institute of Development Studies at Utrecht University. She is conducting research on town–hinterland relationships in Mali and together with Annelet Harts-Broekhuis co-authored the book *Subsistence and Survival in the Sahel*. She is also directing research on changing production circumstances in the Central Province in Cameroon.

Chris Dixon is Professor in the Department of Politics and Modern History at London Guildhall University. Recent publications include: *Uneven development in South East Asia*, Ashgate, Aldershot, 1997 (edited with David Drakakis-Smith) and *Thailand: internationalisation and uneven development*, Routledge 1998.

Annelet Harts-Broekhuis is a geographer with experience in research, training and consultancy work related to development problems in

Africa, focusing on Mali, Cameroon and Zimbabwe. For her doctorate she carried out research in Mali and together with Ali de Jong she co-authored the book *Subsistence and Survival in the Sahel*. At present she is a lecturer at the faculty of Geographical Sciences, Utrecht University.

Anders Närman is Reader in Human and Economic Geography, University of Göteborg, the Director of the Centre for African Studies and Lecturer with the Department for Peace and Development Research (PADRIGU), both at the University of Göteborg. He is also course leader for a graduate course on "International Development Studies", University of Oslo, and "Development and Development Co-operation" at Sida-Sandö. He is also chairman of the Swedish Association for Development Geography (SADG). He has long experience of research in Eastern and southern Africa, mostly dealing with education and general development issues.

Mark Parnwell is Senior Lecturer in South-East Asian Geography in the Centre for South-East Asian Studies, University of Hull. He has specialised in the geography of development in South-East Asia over the last 20 years, having undertaken research and published extensively on a wide diversity of aspects of the development process (migration, rural industrialisation, deforestation, regional development, urbanisation, small-scale enterprises, tourism, China and, especially, Thailand). He speaks two regional languages and has a wide array of institutional connections in the region.

David Simon is Reader in Development Geography and Director of the Centre for Developing Areas Research (CEDAR) at Royal Holloway, University of London. He holds degrees from the University of Cape Town, Reading and Oxford, and has published widely on development theory and development issues relating particularly to Africa, and especially southern Africa. He is also a leading specialist on Namibia. His most recent books are: *Cities, Capital and Development: African cities in the world economy* (London: Belhaven, 1992); (as first editor) *Structurally Adjusted Africa: Poverty, Debt and Basic Needs* (London: Pluto, 1995); Transport and Development in the Third World (London and New York: Routledge, 1996); and (as editor) *South Africa in Southern Africa: reconfiguring the Region* (Oxford: James Currey, Cincinnati: Ohio University Press, and Cape Town: David Philip, 1998).

Rana P.B. Singh, MA, PhD, is presently a senior Reader in Geography, Banaras Hindu University, Varanasi, India, and has been a visiting professor at Virginia Tech (USA), Okayama University (Japan), University of Karlstad (Sweden), and Otago University (NZ). His publications include over one hundred and twenty-six books and anthologies, including *The Layout of Sacred Places* (1993), *Environmental Ethics* (1993) *The Spirit and Power of Place* (1994). As the founding President of an NGO, Society of Heritage Planning & Environmental Planning, he is involved in promoting heritage planning, archaeo-astronomy and pilgrimage & eco-tourism.

Daniel Tevera is a senior lecturer in the Department of Geography at the University of Zimbabwe. He is the co-editor of *Harare: the Growth and Problems of the City* (University of Zimbabwe Publications, 1993) He has published on urban solid waste management, urban poverty and survival strategies, micro-enterprises and several aspects of development. He was a visiting lecturer in the Department of Environmental Science at the University of Botswana for the whole of the 1992/93 academic year.

Ton van Naerssen is Senior Lecturer in Geography of Welfare and Development, in the Faculty of Policy Sciences, University of Nijmegen. His special research interests lie in development theory, urban poverty allocation, with particular reference to southeast Asia. He is also a member of the interdisciplinary Nijmegen Urban Health Group (NUHG), and a consultant for the WHO.

August van Westen is researcher and lecturer at the Institute of Development Studies at Utrecht University. He has conducted research on housing and mobility in Mali. Previously he worked as an advisor on several United Nations development planning projects in the Asian Pacific region. His current research focuses on industrial restructuring and trade in developing countries.

Series Preface

In the late 1980s, the Developing Areas Research Group (DARG) of the then Institute of British Geographers produced a series of three edited student texts under the general editorship of Professor Denis Dwyer (University of Keele). These volumes, focusing on Latin America, Asia and Tropical Africa respectively, were published by Longman and achieved wide circulation, thereby also contributing to the financial security of DARG and enabling it to expand its range of activities. These are now out of print. The Latin American volume was revised and published in a second edition in 1996.

However, there have been dramatic changes in the global political economy, in the nature of development challenges facing individual developing countries and regions, and in debates on development theory over the last decade or so. The current DARG Committee and Matthew Smith, the Geography Commissioning Editor at Addison Wesley Longman, therefore felt that mere updating of the existing texts would not do these circumstances justice or catch the imagination of a new generation of students. Accordingly, we have launched an entirely new series.

This is both different in conception and larger, enabling us to address smaller, more coherent continental or subcontinental regions in greater depth. The organising principles of the current series are that the volumes should be thematic and issue-based rather than having a traditional sectoral focus, and that each volume should integrate perspectives on development theory and practice. The objective is to ensure topicality and clear coherence of the series, while permitting sufficient flexibility for the editors and contributors to each volume to highlight regional specificities. Another important innovation is that the series is being launched in January 1999 by a book devoted entirely to provocative contemporary analyses of Development as Theory and Practice. Edited by David Simon and Anders Närman, this provides a unifying foundation for the regionally focused texts and is designed for use in conjunction with one or more of the regional volumes.

The complete series is expected to include titles on Central America and the Caribbean, Southern and East Africa, West Africa, the Middle East and North Africa, South Asia, Pacific Asia, the transitional economies of central Asia, and Latin America. While the editors and many contributors are DARG members, other expertise – not least from within the respective regions – is being specifically included to provide more diverse perspectives and representativeness. Once again, DARG will benefit substantially from the royalties. In addition, a generous number of copies will be supplied to impoverished higher education institutions in developing countries in exchange for their departmental publications, thereby contributing in a small way to overcoming one pernicious effect of the debt crisis, namely the dearth of new imported literature available to staff and students in those countries.

David Simon
Royal Holloway, University of London

Series Editor
(Chair of DARG 1996–8)

CHAPTER 1

Introduction

Anders Närman and David Simon

From misery to development?

> ... we must embark on a bold new program for making the benefits of our scientific advances and industrial progress available for the improvement and growth of underdeveloped areas. More than half of the world are living in conditions approaching misery. Their food is inadequate. They are victims of disease. Their economic life is primitive and stagnant. Their poverty is a handicap and a threat both to them and to more prosperous areas. For the first time in history, humanity possesses the knowledge and skill to relieve the suffering of these people.
>
> (Point Four, from President Truman's inaugural address in 1949, as quoted by Rist 1997: 71)

This was the confidence with which the rich world embarked on a massive programme of development assistance fifty years ago. Aid was intended to give the poor the impetus to help themselves out of their misery. In a retrospective reflection written shortly before the end of the century, the Development Assistance Committee (DAC), the main aid co-ordinating body of the Organisation for Economic Co-operation and Development, which represents the rich countries of the world, gives a somewhat contradictory picture. While focusing on some of the historic changes taking place in some parts of the Third World, it also notes that

> These impressive strides have not been uniform. In some countries poverty is increasing and in many countries the poor have not shared in the positive global trends described above.
>
> (OECD 1997: 16)

The United Nations Development Programme (UNDP 1997: 2) has also recognised the overall global success of poverty reduction but calls it a scandal and an inexcusable failure of national, as well as international, policies that gross inequalities still exist:

> The resulting disparity and the remaining backlog create and recreate human poverty, a continuing and perpetuating process that the poor constantly struggle to overcome.
>
> *(ibid.:* 48)

To address the challenges ahead, donors are compelled

> to sustain and increase the volume of official development assistance in order to reverse the growing marginalisation of the poor and achieve progress toward realistic goals of human development.
>
> (OECD: 13)

Consequently, from a certain perspective, development assistance has been successful. Some parts of what we have come to call the Third World, or South, and certain social groups have improved their economic status and material well-being considerably (see Chapter 2). On the other hand, this aid has not created a sustainable process of development, built on the combined efforts of donors and the people of the poorer countries. Instead of making itself superfluous, development assistance is now needed more than ever. As will be seen below, new structures of aid dependency have arisen, virtually recolonising some parts of the Third World (see Chapters 2 and 7).

Similarly, development aid programmes have not been able to address the issue of equity, either between nations or within them. On the contrary, the positive processes of change taking place have been followed by the almost automatic exclusion of the many. According to Dube (1988: 114), modernisation has been able to make inequity legitimate, stating that 'around small islands of dazzling affluence there is a cheerless ocean of poverty and degradation'. In actual fact, as pointed out by Cassen (1986), aid is most efficient where it is least needed, and vice versa, from which it follows that aid is least efficient in Africa, at present the world's poorest continent. Under those conditions it seems even more difficult to comprehend the reasoning behind the currently fashionable idea that a neoliberal market economy will change the pattern positively. The World Bank (1994: 220) claims that African pessimism is supposedly unwarranted, as some light is now visible at the end of the tunnel. Nevertheless, this hope wanes somewhat when calculations by the Bank indicate that it will take forty years to bring Africa back to where it was in the mid-1970s! However, 'a sound development strategy and a dose of good luck can change the picture' *(ibid.:* 39). In the meantime, aid will set up safety nets for 'the truly poor or those likely to become impoverished' *(ibid.:* 218). The vision expressed by Truman has been followed by many more, and this process of envisioning development by donors is likely to continue well into the next century. A key question remains whether it is really possible to impose sustainable development for all, even if close co-operation is established among the parties involved.

One of the objectives of this book is to provide a critical reassessment of development thinking and development assistance policy and programmes in the light of current debates about sustainable development and the very mixed results achieved over the last fifty years since President

Truman's famous speech. Although not desirable, it has become conventional for development theory to remain in the realm of academic debate and for development policy and practice to be the preserve of government agencies, NGOs and other development practitioners. While there clearly are exceptions, the deep-seated nature of this divorce has fuelled suspicion and a lack of communication between these different groups. This, in turn, has led to critiques of the 'irrelevance' of development studies as a discipline by practitioners (see below). Added to this, conventional academic practice has hitherto required the almost total depersonalisation of writing in the name of objectivity. For example, it was very much frowned upon until recently to write in anything other than the third person and other than in terms of 'scientific evidence'. Only in recent years have some authors begun to break the mould by writing in the first person, thereby making explicit aspects of their own experiences or positionality. To many, however, this practice remains reprehensible and lacking in rigour. We return to wider issues of subjectivity in development later in this chapter.

Evaluating development

In an overview of development assistance projects, Cassen (1986) offers the opinion that targets and objectives are generally met in a satisfactory manner. However, placed in a broader development perspective, the picture immediately becomes much more gloomy. A normative value-laden assessment of the outcomes of the interventions tends to change the picture substantially. Riddell (1987) makes the point that the complex realities of development cannot be solved by the oversimplified solutions derived from abstract theories. There is no simple or generalised formula for calculating whether aid has achieved desirable results or not. Each case must be assessed according to its particular circumstances. One difficulty in any evaluation of aid is that the available data and comprehensive experiences are often rather limited. A greater emphasis on evaluations – as is already occurring in many cases – could very well form an essential part of the learning process by aid agencies. However, at the same time, decontextualised experiences might not assist us much in future development efforts.

Somehow the positive glimmer of hope achieved by some individual projects and programmes is not necessarily inconsistent with an overall negative outcome (the 'totality' of development). The question that begs an answer is why follow-up evaluations of so many official development projects in the South tend to paint a favourable picture, when the global situation is still characterised by poverty, hunger and underdevelopment. Is it because:

- project evaluation takes place only in relation to the terms of reference set for individual projects, which may themselves be inappropriate?
- individual projects can at best make only a minute partial difference when placed in a broader context?

- various incremental changes, while individually positive, can combine in a counter-productive way, making the total effect reverse direction from that which is intended?
- many development projects are in themselves inappropriate or even negative in their overall impact?
- both symbolically and substantively, 'development' itself means such different things to different people and in different contexts (the latter may even be true for the same individual)?
- development practice, as well as evaluation, is based on some implicit vested interest?

We are faced, in large measure, with a situation in which what is not said might very well be more important than what *is* said. All the positive achievements tend to diminish in the face of gloomy development realities. Possibly the whole truth, as it affects the majority, is ignored in order to continue with an imagined partial improvement for a select group. In a book with the challenging title *The Impossible Aid*, Karlström (1996: 8) has described a dilemma that many of us share – the conflict between the heart and the brain. Too often in practice, the development practitioner finds him/herself trapped in a situation where – for political or other reasons – he/she is unable to do what he/she wants professionally. Alternatively, inspired by a notion of international solidarity, one finds oneself so preoccupied with 'doing good' that one is unable to put the socio-economic effects of a particular aid intervention into their wider context.

Constructions of development

For as long as we can remember, the donor community has claimed to be very active, in terms of refining techniques and methodologies to make aid delivery more efficient. Nevertheless, the principal discussion in the donor community seems to have remained much the same over time; only the jargon and the meanings of key concepts have changed. Whatever is said can be turned into its dialectical opposite, opening the way for alternative interpretations. It is well known that development assistance is a tricky business, under constant scrutiny and critique by the tabloid press and extreme political factions on both the left and right. Donor responses to such attacks often lack substance and focus, as the aim of the adversaries is not to be constructive but to create confusion. With these frequent attacks it is understandable that serious and well-intended attempts to discuss certain development issues more constructively are dismissed in a similar manner.

On the other hand, many academics have great difficulty in explaining themselves in a generally understandable fashion. Concepts and meanings are expressed in as complicated a way as possible instead of trying to simplify matters. This forms part of an academic tradition in the social sciences. An academic discourse is not conducted primarily to create a more in-depth understanding but rather in an attempt to score certain points over an 'opponent'. Furthermore, it is quite obvious that different academic traditions, paradigms or ideologies lead to rather contradictory standpoints. In this we detect hardly any opening for improved practice. Chambers (1997: 33–35) characterises the two cultures of development professionals, i.e. the practitioner and the academic. While the former is preoccupied with action, the latter deals with understanding. Development workers might be too busy solving a single problem, without being able to relate this to complex wider realities, while researchers become too theoretical and ingrained in their own word plays to see the 'reality'.

It could still be assumed that a principal objective of development theory should be to inform development practice, such as development planning and development co-operation. More generally, it would represent an important component in the formulation of development strategies. That this has not always been so seems evident from various experiences. According to Hettne (1995: 11), development studies has never had any substantial political relevance. Coupled with the fact that, in cases where theories have been applied within development strategies, we find a number of failures in practice, this places development researchers in a dilemma. Still, as an academic discipline, development studies has been able to establish a theoretical base for continuous critical debate, but hardly any solutions to the original problems have been found. Instead, new ones have been added to the old ones (see Chapter 2).

From thirty years' experience of engaging with development policy and practitioners within the United Nations, Wolfe (1996: 2–3) adds another dimension to the dichotomy between development practitioners and researchers. His audience, comprising mainly politicians, planners and administrators, generally lacks the interest to discuss development theory. A questioning of the meaning of development might easily rock the boat and threaten international consensus. Instead, these development experts have devoted a chain of conferences to the formulation of declarations that will be translated into development priorities. To critical social scientists who connect the development process to existing conflicts of interest, the search for a harmonious development model often appears to be rather naive and idealistic. Thus the mere identification within development theory of the main national and international agents of development, as in themselves constituting a barrier to development, does not facilitate dialogue between researchers and practitioners.

From our own interactions with aid officials in the course of seminars and lectures as well as development projects, both of us have experienced difficulty in combining theoretical understanding with practical fieldwork. A rather advanced discussion of theory can be followed immediately by distinctly simplistic perceptions of development when turning to consider specific applied examples. Terminology and language use change according to circumstances. The comments by Hettne (1995) and Wolfe (1996) cited above stand in sharp contrast to a recent seminar discussion at the Swedish International Development Co-operation Agency's (Sida's) training centre in Sandö. To a direct question on what most influenced their work, policies and strategies, some representatives of official donors, as well as NGOs, answered confidently that it was a debate about development theory. Paradoxically, this, in itself, might very well be true. However, what practitioners listen to and absorb from an ongoing development discourse might be rather selective. When dealing with theory, the vocabulary and language are totally different from that used when engaging with practical examples of development. That development research could be the instrument by which the people who undertake practical work in the field place their activities into a much broader context is a perception all too often still missing.

Grounding development

This book emerged as a direct result of a seminar for development geographers held at Sida's training centre at Sandö. It was an opportunity for geographers from Scandinavia, the UK and the Netherlands, together with a few invited guests from the Third World, to meet and discuss development issues. This kind of interaction can broaden academic discourse through numerous new influences and ideas. Unfortunately, it is always difficult to find a symmetry between researchers from the South and from Europe, of the sort proposed by Cline-Cole in Chapter 5. To open up our meeting to further dialogue, an invitation was also extended to official Nordic donor organisations and key NGOs. We thereby intended to facilitate an exchange of ideas between the research and donor communities. Development co-operation was supposed to be presented against a backdrop of theory. As such it can be seen as one of a series of potential North–South seminars for academics and practitioners to exchange ideas and concepts in a constructive manner.

For such a purpose, Sandö can be seen both as a perfect environment with a relaxed atmosphere in rural Sweden and as a place holding its own symbolic value. The centre's location was determined within the context of Swedish regional policies, which aim to create some form of equality between various parts of the country. It is this objective, to achieve greater socio-spatial national equity, that is also officially extended across Northern boundaries in the form of development assistance. It is all too obvious that the building of a society based on equity is a complicated matter at both national and international levels.

However, Sandö is much more than that, historically having been a pivotal location in the awakening of a working-class movement in Sweden. When looking at what we in the North can contribute to the South, material hardware is often the most obvious category. To many people in Africa, Asia and Latin America, we represent modernisation and economic success. Still, today's Sweden was built on mobilisation, worker solidarity and awareness. These traditional roots tend to be neglected nowadays, but acknowledgement of them is necessary in any serious North–South dialogue. Verhelst (1990) claims that there is no life without roots, an assertion with which we might agree intellectually in talking about the South. We ourselves must not forget the working-class struggle that took place during the first half of this century along the banks of the river within which Sandö island is situated. It was on that foundation that Third World solidarity in Sweden was built, extending contacts towards

Vietnam during the American military involvement there, to Chile under Allende and to the anti-apartheid movements in southern Africa. In this we can see the origins of what became international development assistance.

An implicit question underlying the foregoing discussion is whether the global situation today would have been better without all the official development assistance contributed since the Second World War. Of course, it is impossible to answer that question in any objective way. Still, it can now be said that perhaps Third World solidarity, built on a tradition of internal mutual assistance among workers, might have been a better starting point for the joint mobilisation of resources for sustainable development than the institutionalised aid that came later. This is so, even if development assistance has been largely influenced by individuals who, at least formally, represent workers' parties or action groups. Leading examples would be Olof Palme in Sweden and Jan Pronk in the Netherlands. Other forms of co-operation could have been established on a model of greater symmetry. Here it is well worth remembering the words of Frantz Fanon (1963: 74):

> During the colonial period the people are called upon to fight against oppression; after national liberation, they are called upon to fight against poverty, illiteracy and underdevelopment.

Alternatives to present trends in development could have been found in an active mobilisation of the population in a joint effort to combat the alien forces of neoliberalism. The example from Chile (Chapter 11) illustrates well the internal dynamics in a country dominated by dictatorship for a long period of time. Early well-meaning attempts to support self-reliance in Tanzania included a sustained programme of aid from Sweden and other progressive donors. The end result was rather painful, however, and indicates that development assistance is a very blunt instrument. However, the Northern predilection for blaming failure squarely on Tanzanian policies and development strategy is paternalistic at the very least; even worse, it could also be a sign of donor hubris. One lesson to be learned from this is the need to be humble and maybe to let the new notion of partnership evolve into some kind of genuine aid reciprocity. At least we have to be able to learn from each other (North–South, horizontally–vertically, academics–practitioners, teachers–students) in an open manner with the intention of creating a real give and take rather than a one-way line of communication.

Subjectivity in development

As mentioned above, a fundamental characteristic of conventional academic practice is that of 'scientific' objectivity, which profoundly depersonalises academic writing. This reinforces many of the unrealistic assumptions that underpin neoclassical economics and its associated development theories – most notably modernisation theory – such as perfect knowledge, competition and divisibility. Perhaps it is therefore not surprising that for many academics, 'development' is principally a matter of dispassionate research and analysis: something 'out there' and external to the self. In other words, it is equivalent to working in any other academic discipline.

However, development *is* fundamentally different. It is about people, many of them far removed from the experience and life world of the academic and hence commonly now referred to as 'distant strangers', even though many of them are actually far closer to home. Moreover, in common with other social sciences, development studies is concerned with promoting change. What distinguishes development studies on the whole is its focus on other, *poorer* territories and societies. This raises a whole set of issues regarding motives, cultural imperialism and ethics (see Chapter 2). It is for this reason that some development authors are now increasingly convinced of the need for a more frank disclosure of personal experience and positionality. This emerges clearly in several chapters in this book, especially those by Parnwell, Cline-Cole, Tevera and Bastías Gonzalez (see also Shrestha 1995; Simon 1998a).

At another level, we are also becoming more aware of 'development subjects' being quite ingenious in sifting the advice and instructions of development professionals. They are thus often far more proactive than we have been led to believe, and are well able to be selective in accordance with their own needs and perceptions through their cultural filters. Far too often, development professionals trained in the North, and/or imbued with 'Western' ways of thinking, take a silence in response to an interview question to mean an absence, a lack of content or knowledge or meaning. Instead, they should understand that silence could arise as a reluctance to reply out of courtesy so as not to disagree or offend, or out of an unwillingness to disclose information deemed private to that individual or his/her social group. For example, Keita *et al.* (1998: 55) express very eloquently how villagers in central Mali revealed their invisible indigenous social institutions and hidden meanings to them (the development professionals) 'only when the villagers were ready for us to know it'. Naturally, this proved a humbling experience.

It also serves as a powerful reminder that even in one of the world's poorest and most marginal countries (in geopolitical, accessibility and agroecological terms), consigned by colonial history to the outer periphery of the world system, local people have not lost their sense of being and their identity. *Francophonie* has not robbed them of their culture or reduced them to subservient and dehumanised native 'others'. They have survived it, adapting to and from it, even if they are now worse off, all against a timeline far longer than the colonial intervention and its aftermath. This is the essence of post-traditionalism in the sense discussed by David Simon in Chapter 2 (and in Simon 1998b).

Many visitors to India over the years have remarked on the dramatic contrast between material poverty and spiritual richness, precisely the opposite to the situation in the 'West'. Westerners and the development orthodoxy automatically assume that material wealth translates into personal contentment, a mindset all too readily assimilated by the middle classes and elites in the South. This is frequently not the case: what seems clear is that modernisation and wealth are commonly achieved at the considerable cost of individualism and spiritual impoverishment. Such dilemmas are also increasingly being faced even in India, where Nehru's modernising legacy is undermining cultural and religious traditions – a challenge taken up by Rana P.B. Singh in Chapter 3. However, the importance of his message is global: it is categorically not a parochial Indian concern.

One illustration of this is Gourley's (1998) short essay on the powerful impact made on her by her first visit to India. As Vice-Chancellor of the University of Natal (South Africa), she was invited to give the graduation address at Visva-Bharati University, established by the famous Indian sage and mystic Tagore in 1921. She quotes Tagore as having believed that 'on each race is laid the duty to keep alight its own lamp of mind as its part in the illumination of the world. To break the lamp of any people is to deprive it of its rightful place in the world festival.' His objective was to extend the university from a purely Indian focus to cover all Asia, in order that Westerners could come to study the great diversity of Asian cultures. At one level, this internationalist vision – as we would call it today – resonates with that of Cecil John Rhodes in endowing the Rhodes Scholarships to Oxford at the height of British imperialism in southern Africa, although with a diametrically opposite motive. To return to Gourley, the power of her experience is interwoven with a deep sense of contrast between Tagore's vision and its embodiment in that university, as in the warm hospitality extended to her, and the generally rather mercenary way in which Northern universities (and those in countries like South Africa) treat foreign students today:

> Attracting foreign students has become in fact a business enterprise. Foreign students are expected to adapt to the local culture and no doubt learn something in the process, but what do we learn from our foreign students? Do we seriously construct our programmes so that local students and staff learn of foreign cultures, practise the hospitality of the mind where the guest is made comfortable by his or her standards and received in a spirit of openness to their culture? It was not always like this.
>
> (Gourley 1998: 20)

In pondering all these issues and subjectivities lies the central objective of this volume, namely to provoke debate and reflection on the many facets of 'development' as both theory and practice as we emerge from the shadow of the Cold War, mark the fiftieth anniversary of Truman's landmark speech and – in Christocentric and secular 'Western' terms at least – anticipate the new millennium.

Structure of the book

The remainder of this book has been divided into two sections of comparable length, addressing development thinking and development co-operation, respectively. Each opens with a broad-ranging essay on the nature of recent changes and current debates, providing an introduction to the more specific chapters that follow.

The first section, entitled 'Rethinking development', comprises five chapters. In this context, it is both significant and appropriate that four of the five authors hail from the South, although David Simon and Reg Cline-Cole now live and work in the UK. The fifth, Michael Parnwell, has many years of experience living and working in Southeast Asia and is married to a Thai woman. The contributors are therefore particularly well placed to analyse debates on cross-cultural issues pertaining to the meanings and implications of development.

David Simon's essay, 'Development revisited: thinking about, practising and teaching development after the Cold War', introduces the section by raising key questions about the way in which 'development' over the last forty to fifty years has been represented by advocates of modernisation as well as by their neo-Marxist and postmodernist critics. He highlights some of the complexities involved in attempting to generalise about the track record of development efforts in different contexts on account of cultural, environmental and other local contingencies. A brief sketch is provided of poststructural perspectives, i.e. postmodernism, postcolonialism and post-traditionalism, and their utility in explaining changing contexts in societies of the South is examined. Finally, he provides a critical review of how and to what extent recent higher education textbooks in English – which inevitably have mostly been written in the North – represent these current debates and issues.

Chapter 3, 'Rethinking development in India: perspective, crisis and prospects', is written by Rana P.B. Singh, one of India's foremost critical geographers. He argues the need for local appropriateness and locally based perspectives in relation to development in India. The predominant tendency to adopt the current Northern fashions uncritically is deprecated, and by way of an alternative, he illustrates exciting possibilities for rethinking the issues and processes in India and beyond that are provided by the wealth of indigenous cultural and religious value systems. Essentially, this is an eloquent and impassioned plea to (re)instate a spiritual dimension into the theory and practice of development.

Michael Parnwell follows this in Chapter 4 by taking issue with the way in which academics within particular social science disciplines tend to deprecate area studies and areal specialisation as second-rate or as purely descriptive local knowledge. He argues that the cultural sensitivity, depth of knowledge and greater ability to sift through the local complexities are vitally important attributes, which can and should be theoretically informed but not be subsumed or lost in the primacy of macro- and often Eurocentric theory, which so many Northern social scientists profess. The chapter is enriched by candid personal experiences, which help to push out the boundaries of reflexivity and debates on positionality.

In Chapter 5, Reg Cline-Cole develops a further critique of current academic practice in relation to development. He provides a critical review of professional practices among human geographers based in the UK and North America and their modes of engagement with, and representations of, Africa and Africanists. He addresses some of the paternalistic and Northern-centric preoccupations and the 'othering' undertaken by many Africanists and discipline-based specialists. Issues of position-ality and misguided, if well-meaning, efforts to bridge the widening North–South gulf that exists in terms of resources and the very different teaching and research contexts are highlighted in this extensive and well-argued essay.

In 'Do they need ivy in Africa? Ruminations of an African geographer trained abroad', Dan Tevera reflects on crossing boundaries, on Northern-centrism in social science higher education and its baggage, and on how appropriately this prepares its initiates to deal with the profound crises and conflicting demands, desires and ambitions that research-active academics face on a daily basis in the South. This provides a compelling individual perspective on the broader issues raised by Reg Cline-Cole.

Section 2, entitled 'Reconstructing development assistance and devel-opment co-operation', evaluates experience with international resource transfers seeking to promote some vision of 'development'. Known suc-cessively as development aid, assistance, co-operation and now increas-ingly as partnership, intervention based on financial, material and technical transfers in order to promote change in the desired direction has been a central tenet of post-Second World War development thinking and practice.

Anders Närman introduces this section with a substantive and broad-ranging discussion of conceptual approaches to, and specific practical policies in, development assistance by Northern donors since the Second World War. Different official and NGO donors are compared and con-trasted, focusing in most depth on evolving policy in Sweden, traditionally regarded as a major and progressive donor (see above). Its current policy thrust is towards forming a development partnership with recipients, with Africa as the 'prototype' region. With often contradictory answers to development questions given by development researchers adhering to various paradigms and ideologies, donors can select freely what kind of practical approach to follow. In spite of the fact that donors often claim to have introduced a new way of thinking, in most cases this amounts only to cosmetic change. Nevertheless, development aid stands to gain sub-stantially from more open debate with the academic community.

In Chapter 8, Ali de Jong, A.C.M. van Westen and E.J.A. Harts-Broekhuis pursue this line of argument with a fascinating study of 'Continuity and change in the Netherlands–Mali bilateral aid relationship'. They explain how, as a relatively small donor, the Netherlands has generally followed the lead of the major multilateral and bilateral donors in terms of policy formulation, although sometimes seeking to humanise the harder ele-ments of aid conditionalities, for example. How Mali, one of the world's poorest countries and a victim of regular droughts in the Sahel, has become one of the key recipients of Dutch bilateral assistance, despite the lack of colonial or other traditional contact between them, makes inter-esting reading. More generally, the case study is instructive of the prob-lems and limitations involved in even relatively enlightened bilateral donor programmes, as well as of some definite achievements.

Chapters 9 and 10 shift the focus to multilateral agencies and donor programmes. First, Chris Dixon argues that the experience of the rapidly growing Pacific Asian economies, so beloved by the International Monetary Fund and World Bank until the dramatic crises of 1997–98, provides a fundamental challenge to these institutions' neoliberal agenda rather than a vindication of it. Following Japan, three waves of economic modernisation can be distinguished, namely the newly industrialising economies in the 1960s and 1970s, the ASEAN Three (Thailand, Indonesia and Malaysia) in the 1980s, and now the transitional economies (especially China and Vietnam) in the 1990s. Their similarities and differences are explored, along with the bizarre somersaults in analysis and attempted explanation of the current crises by the multilateral institutions, which are still desperately seeking to vindicate neoliberal discourse.

In Chapter 10, Ton van Naerssen and Françoise Barten provide a case study of a multilateral assistance programme, namely the World Health Organisation's Healthy Cities initiative. This addresses one of the current policy priority areas, namely the health of large Southern cities and their people under conditions of industrialisation and deteriorating conditions for many residents. Drawing on the broader literature on the subject, preliminary conclusions regarding the achievements and shortcomings of this programme are reached, with reference particularly to a case study of Dar es Salaam.

The final chapter in this section mirrors Section 1 in being a personalised essay based on first-hand experience. María Verónica Bastías Gonzalez, an NGO activist at the helm of Norwegian People's Aid (NPA) in her native Chile, offers critical insights into the workings of a Northern NGO geared to 'people-to-people' aid in Latin America. She focuses particularly on the role of external resources and skills in local politics and grassroots organisation, and the ways in which they are mediated by key actors, often for their own purposes. This is a frank and balanced account, reflecting positive achievements as well as limitations and remaining challenges.

Finally, the editors provide a short concluding chapter, which seeks to draw together some of the major thematic strands of the book. As the introductory volume in a series, this book also serves as a platform for the more detailed regional foci of the remainder of this series.

References

Cassen, R. (1986) *Does Aid Work? Report to an International Task Force.* Oxford University Press, Oxford.

Chambers, R. (1997) *Whose Reality Counts? Putting the First Last.* Intermediate Technology Publications, London.

Dube, S. (1988) *Modernization and Development: The Search for Alternative Paradigms.* UN University, Tokyo, and Zed Books, London.

Fanon, F. (1963) *The Wretched of the Earth.* Penguin, Harmondsworth.

Gourley, B. (1998) World view: Tagore's lamp to the mind, *Times Higher Education Supplement*, 27 February, London.

Hettne, B. (1995) *Development Theory and the Three Worlds* (2nd edn). Longman, Harlow.

Karlström, B. (1996) *Det omöjliga biståndet. Andra Upplagan.* SNS Förlag, Stockholm.

Keita, M.D., Coulibaly, C.O. and **Fredo, D.** (1998) Deep practice: the power of popular imagination and the challenge of doing development differently, *Development* **41**(2): 53–57.

Organisation for Economic Co-operation and Development (1997) *Efforts and Policies of the Members of the Development Assistance Committee Development Co-operation*. DAC Report 1996, Paris.

Riddell, R. (1987) *Foreign Aid Reconsidered*. James Currey and ODI, London.

Rist, G. (1997) *History of Development: From Western Origins to Global Faith*. Zed Books, London.

Shrestha, N. (1995) Becoming a development category, in Crush, J. (ed.) *Power of Development*. Routledge, London.

Simon, D. (1998a) Reflections: writing (on) the region, in Simon, D. (ed.) *South Africa in Southern Africa: Reconfiguring the Region*. James Currey, Oxford; David Philip, Cape Town; Ohio University Press, Athens.

Simon, D. (1998b) Rethinking (post)modernism, postcolonialism and posttraditionalism: South–North perspectives, *Environment and Planning D: Society and Space* **16**(2): 219–245.

United Nations Development Programme (1997) *Human Development Report 1997*. Oxford University Press, New York.

Verhelst, T. (1990) *No Life without Roots – Culture and Development*. Zed Books, London.

Wolfe, M. (1996) *Elusive Development*. Zed Books, London.

World Bank (1994) *Adjustment in Africa: Results, Reforms and the Road Ahead*. Oxford University Press, Washington DC.

Section 1 Rethinking development

CHAPTER 2

Development revisited
Thinking about, practising and teaching development after the Cold War[1]

David Simon

Introduction

This chapter surveys current debates and trends in development thinking, policy and teaching, and seeks to provoke further debate. It offers insights into some of the dramatic paradigm shifts of the last two decades, reflecting in part my own recent rethinking of the potential relevance to the South or Third World of concepts of postmodernism and postcolonialism (see Simon 1998).

In one sense, at least, we currently live and work in an age where anything goes: the certainties and universalising modernising ethos that have characterised mainstream development thought and policy, and which persisted throughout the Cold War, have given way to a flowering of diverse, even contradictory, theories and modes of analysis. While by no means equal or perceived as of equivalent theoretical and practical value, virtually all are at least tolerated in that they have been able to find a particular niche. This apparently postmodern era is commonly characterised as transcending the so-called 'impasse' in development theory which was identified by David Booth (1985) and others in the mid-1980s. This impasse is said to have arisen as a result of widespread disillusionment with conventional development and development failure; the crisis and eventual eclipse of the various strands of socialism as alternative paths; the growing economic diversity of countries within the Third World; increasing concern with the need for environmental sustainability; the increasing assertiveness of voices 'from below'; and the rise of the postmodern challenge to universalising theories and conventional practices of development (Schuurman 1993b).

This book is designed not only to offer insights into recent theoretical developments and reconceptualisations of development and the environ-

[1.] This is an expanded and updated version of a paper published under the title 'Development reconsidered; new directions in development thinking' in *Geografiska Annaler*, **79B**(4), 1997, pp. 183–201. Earlier versions were presented at the Society of South African Geographers' Conference on *Environment and Development in Africa: An Agenda for the 21st Century*, ESKOM Conference Centre, Midrand, South Africa, 29 June–3 July 1997; and the Nordic–British–Dutch Conference on *Rethinking Development and the Role of Development Cooperation*, Sida-SANDÖ, Sweden, 14–17 June 1997.

ment but also, equally importantly, to examine their implications for the practice of development in different contexts. The importance of this is twofold. First, there undoubtedly remains significant scope for improving the nature of interventions made by Northern *and* Southern development workers and agencies, both official and non-governmental. This means enhancing the effectiveness – in terms of specific goals and objectives as well as implementation and monitoring – of both direct interventions and indirect assistance through the provision of funds, for example. It also means challenging the conventional practices and beliefs which serve to perpetuate inequality and the lack of effective (em)power(ment) in the name of humanitarian assistance and political feel-good factors. In this respect, I stress the scope for improvement among Southern as well as Northern institutions and workers, because there is a widespread and, frankly, unhelpful implication in the literature that most if not all of the problems can be blamed on misguided Northern theories and policies. While the simplistic and deterministic constructions of the *dependencistas* have long been discredited, this intellectual legacy remains quite tangible in post- or anti-development and even some strands of postmodern and postcolonial writings. It has, of course, also been reinforced by the strongly negative social impact of structural adjustment and economic recovery programmes and the associated aid conditionalities (e.g. Cornia *et al.* 1992; Woodward 1992; Simon *et al.* 1995; Engberg-Pedersen *et al.* 1996).

The second important reason for rethinking development (in) practice is that important strands of mainstream poststructuralist, postmodern and postcolonial work would have us disengage from practising development at all. Not only is 'the development project' deemed to have failed but the thrust of 'anti-development' writings asserts that it has undermined local vitality and social cohesion. On the other hand, if the implication of the more extreme postmodern challenge to the very basis of collective rationality and identifiable social interest is accepted, then even the possibility of state or NGO interventions in pursuit of 'development' must be illusory and reactionary. The emphasis of much postmodern literature on playful, leisured, heterodoxical self-indulgence also has little to offer those who still can only aspire to safe drinking water, a roof that does not leak and the like. How convenient then to abandon concern, resource allocation and action in the name of fraternal ethical concerns! If this would actually help the approximately 1.2 billion people living in absolute poverty to improve their position, this might be defensible, but I know of no evidence to support such an assertion. This brand of postmodernism certainly would place no more faith in trade than in aid as a vehicle for poverty alleviation.[2] So, unless we are to leave these people

[2.] These critical comments should not be interpreted as a blanket dismissal of the relevance of all strands of postmodernism to the South; indeed, I have recently argued at some length (Simon 1998) that aspects of this perspective do have substantial value in rethinking development and cultural diversity. I shall return to this issue in a later section of this chapter. The point here is simply that extreme cultural relativity militates against any common ground or assumptions – even about the meeting of basic human needs and rights – across societies and cultures. However politically correct in some quarters, this is a negation of common humanity or even modest universalism as regards certain basic values and human rights and is unlikely to be of interest or assistance to those suffering hunger, torture and the like.

and societies to their own devices, to abrogate any responsibility for both distant and often not so distant others, we need to remain concerned with development in practice as much as with development theory (Edwards 1993, 1994; Corbridge 1993b, 1994).

Meanings of 'development'

It is not my intention here to address or even compare the numerous and very different definitions or conceptions of 'development' in the manner of a textbook. These are too well known. For present purposes, it is sufficient to remind ourselves that – at least for even vaguely reflective and reflexive theoreticians and practitioners – definitions are contextual and contingent upon the ideological, epistemological or methodological orientation of their purveyors. Many of these are evident from the labels associated with the multiplicity of approaches to development proffered over the last fifty-odd years by those concerned, for example, with 'reconstruction and development', 'economic development', 'modernisation', 'redistribution with growth', 'dependent development', 'interdependent development', 'meeting basic needs', 'top-down development', 'bottom-up development', 'Another Development', 'autochtonous development', 'autarchic development', 'agropolitan development', 'empowerment', and, most recently, 'post-development', 'anti-development' and even 'postmodern development'.

It is therefore evident that, notwithstanding what some postmodern critics and advocates of mainstream development alike would have us believe (albeit for contrasting reasons), there has never been consensus or unanimity about the meaning or content of 'development'. On the contrary, debate, dissension, contestation and negotiation have been ever-present, both on the ground in particular localities and among the numerous official and unofficial agencies engaged in development work. In an interesting if inaccessible archaeology of 'development', Cowen and Shenton (1996) trace the lineage back to Malthus, Comte, and ecclesiastical writings by J.H. Newman and others in the early decades of the nineteenth century, when it was imbued with spiritual meanings and interwoven with ideas of 'progress', intent to develop and stewardship. However, their frustration at what they see as the incorrect contemporary usages reflects an unwillingness or inability to accept that meanings and usages change and/or are reconstituted over time and in different contexts. By contrast, for example, Leys (1996) is quite explicit about the differences in meanings adopted since the Second World War. Although exploring the postwar process of inventing and elaborating 'development' – and contrary approaches such as 'third worldism' – in fascinating detail, Rist (1997) also emphasises the importance of two pre-Second World War events. These were the hierarchical 'stages of development' drawn up by the League of Nations' mandate system to determine the administrative dispensation for captured territories after the First World War[3], and the

---

[3.] Essentially, a fourfold classification was instituted, known as Classes A–D, which envisaged a rapid process of conferring independence on 'relatively developed' territories but a *de facto* colonial regime for primitive, sparsely populated territories with little modern development.

first major multilateral technical assistance programme, which was provided under the League's auspices to promote China's 'four modernisations' during the 1930s. Escobar (1995) elaborates the use and abuse of development as a vehicle for Western economic and geopolitical imperialism since the Second World War.

Of course, none of the foregoing gainsays the fact that one paradigm, that of modernisation and its contemporary incarnation as neoliberalism, has enjoyed longstanding dominance on account of the power of its institutional advocates and the discrediting of interventionist strategies during the late 1960s and 1970s. If modernisation/neoliberalism has been and remains the orthodox, there has certainly not been a shortage of heterodoxes. We have not had to wait for the postmodern and postcolonial challenges for this. After all, an appreciation of, and challenge to, existing institutional structures, power relations and legitimising propaganda (or discourses, as these are now generally described post-Foucault) have long been central concerns of the approaches generally known collectively as political economy. Lehmann (1997) echoes this point with reference particularly to dependency 'theory'. To a significant extent, anti-development, postmodern and some postcolonial writers have – to modify the parable deployed by Terry McGee (1997) in a recent lecture on a similar theme – set up a straw elephant in seeking to portray the post-Second World War engagements with poverty in the South as a single or singular 'development project' in order to be able to knock it down more easily!

What these recent theoretical 'turns', particularly with their emphasis on deconstruction, have helpfully highlighted and reminded us of is the need for greater self-consciousness, reflexivity and encouragement of difference and heterogeneity (Slater 1997). The inherent and applied value of indigenous traditions, histories and 'knowledges' – especially those lost or marginalised during the colonial and modern(ist) developmentalist periods (the latter often chronologically postcolonial) – have been brought centre-stage as a counterpoint to the often arrogant Westernisation that deprecated or ignored them as ignorant, primitive or simply irrelevant. Nevertheless, it is also important to point out that, especially in those social sciences with strong traditions of fieldwork, including social anthropology and geography, there have always been sensitive researchers articulating the view from below, with the loss of indigenous lifestyles and cultural diversity. Admittedly, this was sometimes sentimental, sometimes Eurocentric and often preservationist in the sense of seeking to create living museums as if behind glass.[4]

However, it can be no accident that it is precisely within these very academic disciplines that the rise of postmodernism has been most hotly debated, and its relevance to the three-quarters of the world's population living in absolute and relative poverty in countries of the South most frequently resisted and denied. Most postmodernists and postcolonialists

[4.] Such museums have now, of course, increasingly become sanitised and idealised as part of the postmodern pastiche experience laid on for international package tourists, although a minority do succeed admirably in their aims of sensitive conservation and recovery of threatened cultures and social identities.

have great difficulty in embracing the *concrete* development aspirations of the poor *in practice*, despite their theoretical sophistication. Part of this trend is a growing retreat to the cosy Northern pavement café – a favoured haunt of those with panoptic vision(s) – from the rigours and challenges of field research in the South, by hiding behind the conveniently hyped 'crisis of representation' of who has a/the right to speak or write on behalf of Third World 'others'. This issue will be returned to below and is also addressed in a somewhat different way by Mike Parnwell in Chapter 4.

All that I wish to add here is that, for me, human development is the process of enhancing individual and collective quality of life in a manner that satisfies basic needs (as a minimum), is environmentally, socially and economically sustainable, and is empowering in the sense that the people concerned have a substantial degree of control over the process through access to the means of accumulating social power. Importantly, although aggregate well-being should be improved as a result, efforts to implement such a development process are almost inevitably conflictual on account of the divergent and contradictory interests of the groups and individuals involved. Given its important qualitative and subjective content, the broad definition just offered naturally defies easy quantification or cardinal measurement. This approach draws on major contributions to the field over the last twenty-five or so years by authors as diverse as Dudley Seers, Paul Streeten, Muhbub ul Haq, John Friedmann and Michael Redclift, perhaps slightly tinged by Wolfgang Sachs and Gustavo Esteva.

Trends in basic needs and quality of life

In assessing the shortcomings or failures of development initiatives, postmodernists and some postcolonial critics downplay or ignore the compelling evidence from around the globe that the dominant aspirations of poor people and their governments remain concerned (albeit for structurally different reasons) with meeting basic needs, enhancing their living standards and emulating advanced industrial countries in some variant of classic modernisation strategies.

Similarly, the very tangible achievements of many 'development' initiatives and programmes – albeit to differing extents and at different rates in rural and urban areas and in almost all countries of the South – in terms of wider access to potable water and increasing literacy rates, average nutritional levels and life expectancy, for example, are often overlooked or ignored. A glance at any recent issue of the *Human Development Report* (UNDP annual) or even the *World Development Report* (World Bank annual), confirms the general trend over the last twenty to thirty years in states of virtually all ideological orientations. A few countries have scored particularly well, as has the Indian state of Kerala, where basic needs improvements have been achieved and absolute poverty has been ended through a locally evolved strategy involving some non-agricultural development and without substantially disturbing the ecological balance. Kerala therefore provides a particular experience of sustainable development that may hold lessons for other states and countries (Parayil 1996).

The principal exceptions to the picture just described are those countries – many of them previously seeking to implement some form of radical socialist programme – where widespread or long civil wars have destroyed physical and social infrastructure and disrupted social programmes. These conflicts were often spawned or fanned by superpower rivalry during the Cold War; examples include Angola, Mozambique and Sudan, El Salvador, Grenada and Nicaragua, and Afghanistan and Cambodia. A more recent and worrying trend towards falling school enrolments (especially at secondary level), literacy levels and access to health-care facilities has emerged since the early 1980s in countries where previously high proportions of state expenditure on education, health and other social services have been severely cut in terms of structural adjustment and economic recovery programmes. Two of the most clear-cut examples are Tanzania and Zimbabwe, as even the World Bank now readily acknowledges in its advocacy of greater attention to the social dimensions of adjustment (e.g. Cornia *et al.* 1992; Woodward 1992; Simon *et al.* 1995; Engberg-Pedersen *et al.* 1996; Husain and Faruqee 1996; UNDP annual).

Overall, the available evidence suggests that, despite debates about how best to implement development, success has at best been uneven, both within and between countries. While average incomes and the quality of life for a substantial proportion of people have been rising over the last two or three decades in much of Southeast Asia and parts of Latin America, for example, the reverse is true in most of Africa, parts of South Asia, the Caribbean and latterly also Central and Eastern Europe. Within many countries, as well as between countries in particular regions, disparities have been widening rather than narrowing, with little evidence that this trend will shortly be reversed in line with predictions of conventional modernisation (e.g. *Africa Confidential* 1997). On the basis of (now somewhat dated) evidence regarding trends within Third World national income distributions as well as at the global scale, Trainer (1996) argues that 'trickle down' has generally been very limited. The poorest 40–60 per cent have made only very small real gains, even when aggregate national incomes have risen substantially. Indeed, the poorest have often experienced a decline in real income under conditions of economic growth. Such evidence is one powerful reason for the growth of so-called anti-development perspectives. Although these critiques of mainstream development(alism) emanate from both the 'post-Left' and New Right, reflecting their respective points of departure, there are actually close similarities between their arguments (see Corbridge 1998). In many respects, this parallels the longstanding similarities in critiques of development aid (more recently restyled 'development assistance' or 'cooperation' – see Chapter 7) from both the hard Right and extreme Left (*cf.* Bauer 1972; Lal 1983; Sachs 1992).

Development discredited[5]

Trainer's (1996) paper is provocatively entitled 'Why development is impossible', and he argues that the obsession with growth as a prerequisite for and indicator of development lies at the heart of the problem. This is underscored by the statistical evidence discussed above. Moreover – and this is a point also made by Nederveen Pieterse (1998) – *dependencia* and other neo-Marxist and structuralist perspectives share modernisation theory's association of growth with development; it is dependent (under)development that they were formulated to criticise. By contrast, according to Trainer, an appropriate development strategy would be more or less indifferent to growth. Such claims are now familiar from the substantial literature on ecodevelopment, deep ecology and related paradigms, although, strangely, Trainer ignores these entirely. Drawing in part upon Davidson's (1992) rather generalised claims that no African peasant societies were thought by European colonialists to have any developmental potential at all, Nabudere (1997) argues that African peasants and other marginalised classes rejected colonial modernisation policies wherever possible on account of their highly regressive impact upon them. Instead, they adopted a two-pronged strategy of *post-traditionalism* (forms of self-reliance and resistance based on their own cultures – see below) and simultaneously dealing with *neo-traditionalism* in the form of modernisation and the institutions invented as 'customary law' by the colonial powers.

Nevertheless, the way in which often diverse programmes, agendas and even principles espoused by very different donor and recipient governments, non-governmental organisations and international financial institutions are dismissed by post- or anti-developmental critics using the fashionable phrase, 'the development project' (e.g. Nederveen Pieterse 1991; Esteva 1992; Sachs 1992; Routledge 1995), is unhelpful, as there neither was nor is such a monolithic or singular construction, even during the heyday of modernisation in the 1960s and early 1970s. Arturo Escobar (1995) exemplifies this trend in a more sustained manner than most by globalising the argument from his penetrating and in-depth analysis of US 'development' interventions in parts of Latin America, especially through USAID. Given his post- or anti-development stance, this rather un-postmodern universalising represents a shortcoming that significantly reduces the power of his critique.[6]

By no means all authors have succumbed to this temptation to universalise: for example, several of the contributors to Crush (1995), especially Porter (1995), Mitchell (1995) and Tapscott (1995), provide nuanced analyses of individual countries, agencies or projects and highlight the interplay between metatheories and broad ideologies, particular discourses and concrete contextual applications. However, many poststructuralist critics

[5.] Parts of this section, like the foregoing subsection, are drawn from my recent analysis of the relevance of postmodernism and related perspectives to the South (Simon 1998).

[6.] I will return to this methodological question later on in relation to other writings on postmodernity and postmodernism.

of conventional development(alism), e.g. contributors to Sachs (1992), still need to take far greater account of the differences in objectives, policy and practice among the various official bilateral and multilateral donors (*cf.* for example, the Nordic countries and the USA; or UNICEF and the World Bank), which were arguably far more substantive during the 1970s and 1980s than in today's neoconservative, market-oriented climate). In addition, many very diverse Northern and Southern non-governmental organisations (NGOs) have adopted very different objectives and methods from official donors over the last twenty or so years, generally working with community-based organisations (CBOs) and so-called social movements, and which have made considerable contributions to both community empowerment and material improvement in quality of life.

Indeed, such organisations and movements do figure centrally in the alternative discourses advocated by Escobar and others; however, the great diversity in every respect of such collectivities and the now-voluminous literature on NGOs warn against idealising them uncritically as embodying the latest 'magic bullet' of development (Walker 1988; Schuurman and Van Naerssen 1989; Ekins 1992; Schuurman 1993a; Edwards and Hulme 1995; Hudock 1995). Somewhat bizarrely, given his trenchant and detailed critique of official discourses and development policies and programmes, Escobar (1995) adopts a sharply contrasting and ingenuous idealisation of NGOs and 'new' social movements as authentic and legitimate, without any attempt at evaluation or deconstruction. David Lehmann (1997) has recently underlined this latter point most forcefully within a wider critique of Escobar's book. The contrast between Escobar's treatment and Schuurman's equally theoretically informed discussion of NGOs is sharp. Jan Nederveen Pieterse has also shifted ground substantially from his (1991) position, cited above; he now adopts a position very close to my own, arguing that post-development thinking in the 1990s generally simplifies 'mainstream development' unjustifiably into a single, homogeneous thrust towards modernisation (1998: 347).

Conversely, developmentalists all too often still ignore or fail adequately to internalise the reasons for widespread 'development failure', especially in poor countries and among often large subordinate, unpowerful groups, and therefore the potential value of postmodern, postcolonial and related visualisations. It is indeed ironic that the absolute or relative failure of many developmentalist states and state-led development strategies is central to neoliberal and post- or anti-development approaches alike. Hence, the rolling back of the (generally developmentalist) state, one of the central tenets of current neoliberal development orthodoxy, is increasing the political and symbolic spaces for, and hastening the evolution of, diverse NGO and CBO initiatives in many countries,[7] which some writers see as constitutive of postmodernity (Bell 1992; Escobar 1995).

---

7. For various reasons, it is by no means inevitable, however, that political liberalisation and the establishment of formal multi-party democratic structures will translate into greater freedom and diversity of practice within civil society and NGOs (Hudock 1995; Simon 1995a: 35–41).

Simply to dismiss postmodernism and related paradigms as irrelevant or esoteric without any attempt at serious evaluation of, or engagement with, them is both methodologically and practically unhelpful (see below).

The rapidly expanding literatures on globalisation and 'flexible' post-Fordist production in the world economy have been quite successful, albeit unevenly, in examining the interconnectedness of the divergent economic fortunes of different countries and regions across the globe (see recent reviews by Barff 1995; Thrift 1995; Whatmore 1995). However, these latter perspectives are still dominated unequivocally by Northern-centric world views. Little consideration is given to possible alternative perspectives focusing on local world views and development strategies or ideologies that rely rather less on external determinants (e.g. Adedeji 1993; Himmelstrand *et al.* 1994; Barratt Brown 1997; Culpeper and McAskie 1997).

Many of the contributors to these books share Afrocentric world views, are critical of the inequities of the existing world order and the colonial or neocolonial relations that have given rise to the current crisis of sustainability. Yet their perspectives would not be considered as 'postcolonial' by adherents of that paradigm. Indeed, the diffuse literature on postcolonialism connects remarkably little with conventional developmental agendas or – far more surprisingly – with postmodernism, despite the former's valuable focus on restructuring inequitable colonial inheritances, and the cultural politics of identity, especially recovering the 'lost' identities of groups subordinated and marginalised by colonial practices, official histories and Northern feminist and environmentalist discourses. This fragmentation of discourse, or perhaps more accurately the politics of discourse and labelling, will be returned to below.

In view of their very different points of departure and agendas, it is somewhat ironic that the two dominant occidental development paradigms of recent decades, namely modernisation and political economy/ structuralism, have generally shared the characteristics of being rather narrow, often economistic, top-down and overtly modernising in application. They also share the characteristic of being overarching metatheories, firmly rooted in the discourses of intellectual modernism, and therefore seeking to provide singular, universal explanations for poverty and underdevelopment and prescriptions for overcoming them.[8] However, it is worth reminding ourselves that modern development is not a totally uniform or smooth process, and that modernisation need not lead to global homogeneity, especially if undertaken with a degree of politico-economic and cultural autonomy, as the Japanese experience illustrates so powerfully.

The objectives of conventional developmentalism with respect to the South are generally articulated at three principal levels, although the particular discourses, agendas and processes of development may differ considerably both within and between them:

- by the populations of poorer countries, expressed, for example, in voting patterns. This can be illustrated by Peru's President Fujimori

---

[8.] This point has recently been elaborated more fully by Taylor (1996).

winning widespread popular support in that country's 1992 general and presidential elections by virtue of his relative success in clamping down on Maoist guerrillas and his promises of better living standards, despite the undemocratic route by which he had seized power a few years previously. More generally, the struggles by poor people to meet their basic needs and their aspirations for an improved quality of life are strongly influenced by the demonstration effects of modernisation and the consumerist lifestyles of the middle and upper classes. Different methods of and routes to achieving these goals may be adopted, but active alienation, rejection and rebellion are normally only last resorts.

- by nation-states, in terms of their political programmes and national development plans. For example, the Zimbabwean government has consistently sought to prevent and eliminate squatting and informal urban settlement on the grounds that it is demeaning and unworthy of a progressive, modern (and until recently also supposedly socialist) African state. Depending on the nature of the state and openness of the political system, regional and local state institutions may share or oppose the central state's agendas, but extreme measures such as active rebellion and attempts at secession may be increasing in frequency as the writ of ossified and corrupt highly centralised states faces challenges from outlying and impoverished areas. The armed resistance by the Sudan People's Liberation Army and other Christian and animist groups to Islamicisation in southern Sudan by the government in Khartoum, the Zapatista rebellion in Mexico's Chiapas state, the insurgencies against Mobutu Sese Seko's kleptocratic former dictatorship in Zaire (now the Democratic Republic of Congo) and the Karen hill people's struggle against the brutal military regime in Burma (Myanmar) are all cases in point.

- by international financial institutions (IFIs) and donor agencies, in terms of their overarching discourses, lending criteria and funding priorities. For instance, the World Bank and other donor agencies have continued to promote large-scale dam projects and other infrastructural programmes in order to maximise conventional economic benefits, despite the well-known social and environmental costs and evidence that smaller schemes, built with greater sensitivity to local people and their environment, are often also economically more successful. In fairness, some greater attention has been devoted to the social and environmental consequences of large schemes in recent years, but – with one or two notable exceptions – generally still predicated on the assumption that construction should go ahead, e.g. the Narmada River dams in India, the Turkwel Gorge dam in Kenya's West Pokot district and the Three Gorges Dam on the Yangtze River in China.

More emphasis by donors and recipient governments on alternative delivery systems,[9] processes and project types emerged during the 1980s,

---

[9.] This term is still in wide use among development agencies but is itself problematic in contemporary terms, as it implies top-down provision. However, much of the thrust of recent changes to the practices of such agencies has been concerned with becoming more locally responsive and involved collaboratively – either directly or indirectly – in the process.

not least because of funding constraints and conditionalities, themselves linked to the new deity of economic efficiency and marketisation. However, such co-option often devalued more radical alternative antecedents, reducing them from agendas for change and empowerment into little more than shopping lists that are hawked to donors for implementation, commonly more in line with donors' than recipients' priorities (see also Närman, Chapter 7). This has been particularly graphically illustrated with respect to basic needs philosophies (*cf.* Wisner 1988; Bell 1992; Simon *et al.* 1995; Streeten 1995; Wolfe 1996) and, I would argue, is currently being repeated in relation to the ubiquitious sloganising about 'sustainable development'.

As I argued some years ago, the pedigree of the sustainability debate stretches back at least to the early 1970s (Simon 1989), when the impact of Rachel Carson's (1962) landmark catalogue of development's environmental woes in the USA, *Silent Spring*, and neo-Malthusian concerns about resource exhaustion, prompted important new research agendas, major international conferences and the establishment of the United Nations Environment Programme (UNEP). Certainly, today's environmental discourses are very different from those evident in *The Limits to Growth* (Meadows *et al.* 1972), *A Blueprint for Survival* (*The Ecologist* 1972), *Only One Earth* (Ward and Dubos 1972), *Small is Beautiful* (Schumacher 1973) or *The Social Limits to Growth* (Hirsch 1977) and indeed from the development agendas informing 'Reshaping the International Order' or the Brandt Commission reports. However, what at that time was still widely regarded as a radical or eccentric fringe concern has become progressively more accepted and acceptable over the intervening years. The establishment of the World Commission on Environment and Development (WCED) (the Brundtland Commission) in 1983 and the publication of its report, *Our Common Future*, in 1987 both reflected this and provided a new landmark in the 'foregrounding' of sustainable development as discourse, objective, process and fad. In the same year, Michael Redclift's (1987) elegant little book, *Sustainable Development*, appeared, taking conceptual and analytical rigour in the field to a far more sophisticated level.

Five years later, the WCED report had a sequel in the form of the 1992 UN Conference on Environment and Development (UNCED) in Rio de Janeiro, intended to transform the concept into more concrete international commitments and agendas. Notwithstanding substantial horse-trading and the watering down of the intended conventions for ideological and domestic political reasons by the US (Republican) and British (Conservative) governments in particular, a process that generated much criticism and cynicism among many environmentalists and radical NGO critics, the conference did result in unprecedented intergovernmental and NGO commitments to biodiversity conservation, greenhouse gas emission reductions and the implementation of Agenda 21 at local as well as national and international levels (Middleton *et al.* 1993). Subsequently, there has been widespread evidence of greater flexibility and commitment to more diverse project types and scales and to greater environmental prioritisation in various policy arenas (Hurrell 1995; Reed 1992). Nevertheless, it is also certainly true that most official agendas en-

Figure 2.1   A cartoonist's view of the 'Rio Plus Five' summit in New York, 1997 (Zapiro in the *Mail & Guardian*, 27 June 1997)

visage little fundamental change, focusing on promoting more efficient resource and energy valuation and use, recycling and reduced pollution broadly within existing parameters rather than on radical changes to lifestyles and economic systems. At the extreme, sustainable development has become a convenient slogan to signal political correctness without the corresponding commitment to change. Such expediency is usually associated with establishment institutions, as was once again underlined by the failure of the 'Rio Plus Five' summit at UN Headquarters in New York in late June 1997. Many government ministers and NGO activists alike were very critical of the lack of commitment by key Northern governments, and dubbed the event 'Rio Minus Five' (*Independent* 28/6/97). A critical cartoon in the South African weekly independent newspaper, the *Mail & Guardian* (27/6/97), pictures a choking Earth reading a news report on the broken promises over global warming being discussed at the summit and exclaiming, 'Just what I need ... more hot air!' (Figure 2.1). Perhaps similar cynicism prompted Escobar's (1995: 192–193) condescendingly dismissive assertions about sustainable development and the Brundtland Commission report, to which David Lehmann (1997: 574–575) has justifiably taken such exception.

The dominant modernist developmental ethos is still for the most part obsessed by the agenda of economic efficiency, articulated largely through privatisation and liberalisation programmes. These programmes have long pedigrees but derive their immediate impetus from aid conditionalities imposed by the IFIs and other donor organisations as strategies for overcoming the Third World debt crisis and promoting 'free' international trade. The logical – and, indeed, desired – outcome has been the almost universal rolling back of the state coupled with a resurgent role for domestic and especially international capital, even in peripheral post-socialist states (Hanlon 1991, 1996; Sidaway and Power 1995). The ultimate prescription has been to maximise trade through export-oriented production based on supposed international comparative advantage.

While this approach may improve the delivery of certain goods and services, it generally and deliberately fails to address power and equity issues adequately and is likely to undermine the ability of developmental states to deliver on their political programmes for social development, from which their legitimacy has been sought and derived. Moreover, there is a deep-seated tension between the cutting of social expenditures in line with donor conditionalities (despite some more recent palliative packages to address the social dimensions of adjustment) and the promotion of literate, healthy and active participants in expanding democratic structures and civil society (e.g. Simon *et al.* 1995). In effect, it also has to be realised that, particularly in their earlier 1980s formulations but also more recently, conventional analyses of the debt crisis and the most effective solutions amounted to blaming the *victims* of development, the vast majority of whom had little if any say in the policies adopted by their states or the transnational banks and other financial institutions and official donor bodies. This is closely linked to 'Afro-pessimism' and its equivalents in other regions. Similarly, SAPs and conditionalities have been described by their prescribing doctors as harsh medicine required to effect a systemic cure. Yet, like most conventional Western medicine, they are directed at the symptoms rather than the underlying causes (Simon 1995b).

In terms of the prevailing conventional economic development wisdom, greater market orientation would actually enhance the prospects for attaining modernity by achieving economic growth, which is widely regarded as being an essential prerequisite for subsequent redistribution and the wider fulfilment of basic needs and popular aspirations (Slater 1993, 1995b). The one dimension of equity that has generally received increasing attention is that of gender, awareness of which is now accorded explicit recognition in most policy and programme documents, albeit still frequently merely at the level of lip service or superficiality in the 'women in development' mould. More thorough-going integration of gender issues in accordance with 'gender and development' and 'gender relations' approaches is still inadequate *in practice*, despite the now increasingly prominent position of various feminist discourses in development debates, especially around indigenous rights and identities as well as community participation and the environment. Promoting such gender integration is, of course, quite different from the particular radical brand of feminism as an intellectual or research paradigm (Bell 1994; Minh-ha 1989; Moser 1993; Nesmith and Radcliffe 1993; Radcliffe and Westwood 1993; Shiva 1988; Townsend 1995; Marchand and Parpart 1995; McFadden 1997).

Post-everything

The current theoretical 'turns' are characterised by the prefix 'post-' in relation to most periods or paradigms, as in postcolonial, postmodern, post-Cold War, post-development and so forth. Clearly they are used to signify differences, either in terms of periodisation or conceptual and methodological approaches. We could, therefore, be forgiven for suffering a degree of 'post-itis', of feeling past it, post-everything! After all, even history has ended, if Francis Fukuyama's simplistic triumphalist credo is to be believed. In a similar vein, some recent discussions about time–space compression in the context of globalisation and the role of telecommunications have suggested the end of geography, as if space had somehow triumphed over place in the sense of localities being imbued with specific socio-cultural meanings.

What I am suggesting is the importance of a healthy scepticism towards some of the more sweeping and emotive formulations of post-everything, which may universalise from particular case studies in a manner reminiscent of modernist theorising, be elitist as practised by its advocates despite the supposed concern with precisely the opposite, and may actually be of little practical use in addressing poverty and providing basic needs. Moreover, critiques of conventional developmentalism and the search for more meaningful, appropriate and socially grounded and bottom-up alternatives are not new. As with the different definitions of development and the examples of basic needs and environmental sustainability given above, there is a long pedigree of initiatives and theoretical formulations stretching back decades and including, for example, Reshaping the International Order (RIO); autarchy as advocated by the extreme dependency authors, André Gunder Frank and Samir Amin; the Brandt Commission; Another Development – as articulated by the Dag Hammarskjöld Foundation through its journal, *Development Dialogue*, since the late 1970s (e.g. *Development Dialogue* 1978, 1980); the agenda for a New International Division of Labour articulated through UNCTAD; and a range of grassroots and bottom-up strategies from different 'alternative' perspectives, of which agropolitan development, associated with John Friedmann, is possibly one of the best known. Nederveen Pieterse (1998) also argues that there is no clearly identifiable 'alternative development paradigm': this term has come to embrace a wide variety of different positions and critiques, many of which concern the 'how to's of development rather than the nature of development as such' (*ibid.*: 352). The same applies to post-development, which also

> parallels postmodernism both in its acute intuitions and in being directionless in the end, as a consequence of the refusal to, or lack of interest in translating critique into construction.
>
> (*ibid.*: 361)

He ultimately prefers a form of 'reflexive development' in which a critique of science forms part of development politics. Rosemary Galli (1992) has also examined anti-development perspectives that have little to do with postmodern or postcolonial critiques. The same is true of the 'autonomous development' envisaged by the contributors to Culpeper and McAskie (1997).

That said, and as I have recently argued at length (Simon 1998), it is no longer appropriate to reject these 'post-' perspectives out of hand as being irrelevant to societies in the South. Many of the problems and non-debates have arisen from imprecisions in the use of the terms 'postmodern' and 'postcolonial'. I discuss a threefold distinction between the postmodern as period or epoch, as mode of expression or aesthetic form and as analytical method or problematic (ways of seeing), which is very helpful in disentangling the range of usages, and the last of these three has most potential in relation to the South. I then suggest the application of this same threefold categorisation to the literature on postcolonialism; although it is somewhat more difficult to separate them, it is again the postcolonial problematic that appears to have the most utility. I shall now briefly explain why.

Postmodern perspectives

What distinguishes the present period is that the expression of conventional developmental ideals and the methods of implementing them no longer enjoy universal acceptance and legitimacy within targeted countries and areas. Increasingly, individuals and groups of people at a local level are either seeking the attainment of their aspirations for better living standards outside the realm of the state or they have rejected the dominant developmental discourse(s) and are pursuing alternative agendas with very different aims and objectives. In the former case, they are still seeking the basic needs and other fruits of modernisation but have despaired of the ability of the state and official development agencies to deliver on their promises, and have thus taken their own initiatives. In the latter scenario, they have rejected the basic premises and trajectories of the modern developmental state. Hence, urban and other 'new' social movements have arisen in a wide variety of contexts and countries in response to a vacuum or, more generally, as alternative modes of organisation and with very different agendas from discredited official local government or community structures (Walker 1988; Schuurman and Van Naerssen 1989; Routledge 1993; Bell 1994; Edwards and Hulme 1995; Hudock 1995). Social and environmental dimensions of protest and action have been linked (Schuurman and Van Naerssen 1989; Schuurman 1993; Radcliffe and Westwood 1996).[10] A dramatic recent example, which integrates development and environmental concerns, is the citizen's rebellion in the Mexican town of Tepotzlán in late 1996, when the mayor and town council were expelled and a virtual unilateral declaration of inde-

[10.] This linking of different 'arenas' of concern also highlights important differences between narrow Northern legal definitions of human rights and initiatives to universalise them through UN Conventions and the like, and broader Southern approaches, often derived from indigenous systems, where economic, political and environmental rights are no less important. Recent debates on the issue have often become embroiled in the postmodernist/modernist polemics of cultural relativism/universalism. Suggestions for transcending this impasse include a greater willingness to learn from non-Western contexts, as principles and practices – often derived from 'traditional' roots – in the South may prove appropriate in the North (Penna and Campbell 1998), and attempts to define some acceptable minimal universalism (Corbridge 1998).

pendence was issued over the mayor's efforts to have a major US$400 million upmarket development comprising a golf course, other sports facilities, a hotel and condominium of 800 homes in the name of 'development', while ignoring popular development demands. In addition, the golf course would have exacerbated the local water shortage and put valuable land beyond the reach of most residents:

> In Tepotzlán, however, where cars must squeeze into cobbled streets meant for donkeys, ... the local residents were not buying the golf club's passport to modern life. Petitions demanding the cancellation of the golf club turned into street protests, and then into demonstrations outside the town hall. When Morales [the mayor] still refused to meet his angry constituents, a group stormed into his offices and held six officials hostage. The rebellion had begun. ... 'It began as an environmental protest,' says Rodriguez [the protest leader], ... 'but with the jailing of four comrades over the past year, and two deaths in clashes, and all the arrest warrants hanging over our heads, it has become much more complicated. We cannot give up the fight now.' ... 'A unique and extraordinary phenomenon is taking place in Tepotzlán,' Adolfo Aguilar Zinser, an opposition congressman and longtime resident, wrote in the daily La Reforma. 'We, the residents of Tepotzlán, are discovering that no government is better than bad government. Without a PRI government, without municipal police, without the presence of any federal law enforcement agency, we enjoy a far higher level of security than in the rest of the state of Morelos.'
>
> Not everyone shares Aguilar Zinser's rose-tinted views. Some residents say the town has become more polarised, while many are tired of the endless appeals for money to keep the rebel government afloat. Relatives of ousted officials who remained in Tepotzlán have suffered discrimination and abuse.
>
> (Crawford 1996)

This example illustrates well how the politics of local protest, induced by popular rejection of conventional development agendas that are perceived to be imposed in a top-down manner by unresponsive elected officials and developers, can, if the sentiments are deeply enough felt and the authorities sufficiently inflexible, progress to more direct action in defence of space, place and popular aspirations. The outcome was apparently unforeseen and unimagined by any of the protagonists, but the stakes were raised and the result was open rebellion and the usurpation of the local state by the protesters and their supporters. No doubt there were other local complexities, and the article says nothing about the socio-economic profile of the activists or community at large. However, the writ of the hitherto omnipresent PRI no longer runs in Tepotzlán. Whether the stand-off will persist and whether the residents will be able to organise an alternative system of local administration remains to be seen. However, this example highlights the importance of contingency and locality in the analysis of events and movements for change, even in this age of globalised communications and 'glocalised' consciousnesses and identities. Similar processes could be analysed in respect of the conflict in Somalia and the survival of the breakaway component in Somaliland.

A very different example is provided by the response of Sebastian Kamangwa, headman of the 4,000-strong Shitemo community living in an isolated district of Okavango region in northeastern Namibia, to the recent opening there of a primary health-care clinic by the country's minister of health. At a time when conventional and traditional (bio)medical systems are increasingly coming together in complementary syntheses (which are arguably postmodern – see Simon 1998) in various parts of sub-Saharan Africa and beyond, he reportedly proclaimed categorically that:

> In the past malaria caused a lot of suffering and fighting because people thought it was the result of witchcraft, but now we have seen that the clinics can solve these problems. Some people are still trying to cause trouble by demanding that traditional healers be revered, but I am adamant that we cannot have traditional healers working alongside modern health services.
>
> (*The Namibian* 6/6/97: 8)

Such an overtly modernist stance might seem rather outdated or even quaint, yet the headman clearly perceives himself as progressive. This exemplifies my earlier points about social conditioning – modern or otherwise – and the powerful demonstration effect, albeit substantially time-lagged, of perceived successful modern innovations in otherwise apparently conservative rural communities. It also raises several questions about representation and legitimacy within local communities; in other words, how representative is the headman's stance of his people's perceptions, and will their practices change in view of his attitude? There is also a major issue for social theorists to ponder. The new clinic saves people a walk of at least 12 km and helps to treat malaria and other serious illnesses. Is it therefore legitimate for postmodern and/or post-colonial critics to decry or dismiss the significance of such innovations to poor people's lives? I return to this question below.

These contrasting examples also demonstrate that there is a growing disjuncture between modernist developmental rhetoric and the increasingly diverse experiences of such programmes on the ground. While spokespeople, political leaders and even many 'grassroots' or community groups remain committed to the grand scale and 'the big ideas' of progress and development, methods of implementation invoke strategies that, elsewhere in the world, have been associated with the postmodern.

What is therefore emerging is a growing acceptance of heterodoxes, diversities and multiple systems, explanations and modes/scales of institutional organisation, which are at least partially superseding the conventional modernist traditions of a single orthodoxy in state ideology and practice. However, it is by no means certain, or even desirable, that this trend will eventually eliminate modern(ist) development agendas. Therefore, it may well be that the co-existence and simultaneity of diverse (and even divergent) systems and practices become an enduring reality, even though their relationships are likely to be flexible and changeable, and perhaps as much symbolic as substantive. This condition exemplifies the essence of postmodernity as understood by analysts working in the North (Bauman 1992; Berg 1993; Dear 1988; Folch-Serra 1989; Featherstone 1991, 1995; Harvey 1989; Soja 1991; Watson and Gibson 1995), in terms of which the monolithic modernist discourses, both liberal and Marxist, have been or are being discarded in favour of a multiplicity of ideologies and modes of explanation.[11] In terms of the schema discussed above, this represents the notion of the postmodern as problematic, overlain with a distinct element of the postmodern as epoch, albeit *without* a clear break from the modern and, indeed, characterised by the co-existence of and overlap between the two.[12]

In many respects, this conceptualisation appears to offer a far more helpful way of understanding the often disjointed and conflicting processes, phenomena and material and cultural styles – both urban and rural – that are now so typical within countries of the South as well as a way to help to rethink North–South relations (e.g. Slater 1992a, 1992b, 1995a, 1997). This will be evident to anyone who has encountered the jarring contrasts on stepping out of an ultra- or postmodern urban shopping precinct into untarred streets lined with shanties and beggars, or who has encountered the paradoxes of contemporary tourist landscapes superimposed on poor rural communities in the Caribbean, Latin America, Africa or the Asia-Pacific regions. Indeed, it may well be that this condition is far more widespread and characteristic of the South than the North. It is also not necessarily a very new or recent phenomenon – having roots at least as far back as the late colonial period in Africa, Asia, the Caribbean Basin and Pacific Islands – but rather a different way of seeing and interpreting the quite longstanding phenomena of Southern dislocation, unemployment and poverty, previously regarded as representing incomplete modernisation and the iniquities of colonialism. Moreover, many of the

[11.] However, it is important to reiterate that, as explained earlier, the implicit globalising and universalising of 'the postmodern condition' by authors like Harvey and Soja reflects a continuing modernist methodological praxis. Featherstone (1995: 79) has recently amplified this point in relation to the overall methodology of Harvey and Frederic Jameson, namely their conception of postmodernism as a cultural form accompanying the transition to late capitalism or flexible accumulation:

> Like Jameson, Harvey sees postmodernism as a negative cultural development with its fragmentation and replacement of ethics by aesthetics leading to a loss of the critical edge and political involvement which he regards as characteristic of the works of artistic modernism. Yet ... [such] analyses rely on a totalizing logic which assumes that the universal structural principles of human development have been discovered and that culture is still caused by, and is a reflection of, economic changes. ... They rely upon a neo-Marxist metanarrative and metatheory which insufficiently analyses its own conditions and status as a discourse and practice. ... This leads to an inability to see culture and aesthetic form as practices in which their meanings are negotiated by users. It also displays an inability to see that economics should itself be regarded as practices which depend upon representations and need to be seen as constituted in and through culture too.

[12.] Naturally, different forms and processes may occur in different regions and contexts. For example, Leontidou (1993) has argued that postmodernity in southern Europe has been characterised by a transition from preindustrial to postindustrial without having experienced widespread industrialisation. This challenges the conventional linear stage approach to development as well as views of the postmodern as being strictly epochal. Colás (1994) engages with a variety of literary and cultural perspectives from Latin America.

contrasts, contradictions and fragmentations of meanings and practice within the South are at least as much the result of deliberate or wilful actions as is the case with postmodern showpieces of urban design and other forms of expression in the North.

Postcolonialism – a Eurocentric construct?

Dani Nabudere, the veteran radical Ugandan lawyer, social scientist and politician, takes issue with the entire notion of postcolonialism (personal communication 14/1/96). He regards this as too Eurocentric, implying both the previous hegemony of colonial institutions, social structures and identities as so eloquently elaborated by Blaut (1993) and Corbridge (1993a), and that the experience of colonialism is the defining point of reference. However, in many parts of the former colonial world, including sub-Saharan Africa, indigenous values, social structures and identities survived – admittedly to differing extents and with differing degrees of engagement with or transformation by colonial impositions. Hence, in his view, the task of evolving and promoting new, people-centred and indigenously generated African alternatives to the colonial and the modern should be more accurately termed 'post-traditional' (see also Nabudere 1997).

A fascinating example is provided by the landmark investiture of Sinqobile Mabhena, a young female trainee primary school teacher, as chief of the 100,000 Nswazi people (an Ndebele subgroup) in Zimbabwe in December 1996. Under the headline, 'The chief who wears a miniskirt', Andrew Meldrum wrote thus in *The Guardian* (24/12/96):

> Surrounded by government ministers and tribal chiefs, Sinqobile Mabhena appears a model of female subservience as she bows her head and modestly lowers her eyes. But this demure 23-year-old has rocked Zimbabwe's traditional culture by becoming one of the first women to take on the powerful mantle of tribal chief. "I know many people are opposed to me becoming chief because I am a woman" said Ms Mabhena. "Being a woman doesn't mean you are disabled." ...
>
> "Chief Howard Mabhena died in 1993 and he had no son to succeed him. It therefore fell upon his eldest daughter, Sinqobile, to succeed him and she has taken up her chieftainship responsibilities with humility" [said John Nkomo, minister for local government]. Ms Mabhena's investiture has been delayed by more than a year because of objections. "The government held lengthy discussions and the Nswazi people insisted they would rather have Sinqobile than a male chief who would not have been appointed by them. All's well that ends well." ...
>
> Mr. Nkomo invested Ms Mabhena with the traditional chief's costume, an incongruous but arresting mix of African and colonial symbols of authority: a crimson and purple chief's robe, a white pith helmet, a leopard skin and a staff. ...
>
> "I want to look at all sides in any dispute and to be fair," said Ms Mabhena. "I don't want to only take the woman's side or to just take

the man's side." … As scores of well-wishers crowded around to congratulate Ms Mabhena, the young chief wiped a few tears from her eyes. "I just thought about this whole thing, the history, my father, the future, the responsibility, everything," she said. Sinqobile Mabhena is a combination of the old Africa and the new. During the week she lives in Bulawayo, where she is studying to be a primary school teacher. She wears short skirts, high heels and has a boyfriend, who is a schoolteacher. At weekends she goes back to her family's rural home where she meets in council with the Mabhena clan's elders.

Her investiture, however, does not bring an end to the controversy. The ceremony was boycotted by several chiefs and political figures and the attendance of 800 was smaller than the 2,000 that had been anticipated. "I remain opposed to this because it is against our culture," said Welshman Mabhena, governor of the neighbouring Matabeleland North province, who did not attend the investiture. "An Ndebele chief must always be a man."

George Moyo, chairman of the Vukani Mahlabezulu Cultural Society, also opposes the investiture of Ms Mabhena. "Our ancestors did not approve of a female chief. It is going to destroy our culture. In our culture women were only advisors at home, that's all," he said. "There are many chiefs who are not going to accept this. The Nswazi people are going to have trouble because of this."

Ms Mabhena's grandmother, Gogo Flora Masuku, is outspokenly in favour. "I am very, very happy to see a female chief. Women must stand up for their rights and advance their position. Women fought to end Rhodesia. We now have female cabinet ministers and airplane pilots. Why not chiefs? Is the queen of Britain a man? Is Margaret Thatcher a man? Women can be leaders."

Here we have a rich tapestry of cross-cutting continuity and change, of old, new and hybrid identities, of reason and reaction, of gender and power relations, of the preservation versus transcendance of categories, and of how and by whom they are negotiated, defined and safeguarded. These issues are all chronologically and analytically postcolonial. Moreover, we have here a timespan and a problematic that simultaneously engages the indigenous, precolonial 'traditions'; the colonial institutions and laws that subordinated indigenous practice into a category of 'customary law' and its upholders, embodied in the office of chief, into clients of the colonial state; and the ongoing struggles and challenges of the postcolonial epoch, one critical dimension of which is the relationship between the state and indigenous institutions. We are given the broadly accurate impression of a powerful government which nevertheless treads warily with regard to custom and chiefly powers: it sought a mediatory rather than a prescriptive role, yet ultimately reaffirmed its statutory and effective primacy over the customary realm by despatching no lesser a representative than the minister of local government to install the new chief. The ironies are considerable, ranging from the quintessentially colonial name of the one male traditionalist (i.e. Welshman) seeking to uphold precolonial norms, and the Scottish ('European') first name of the feminist grandmother, who invokes examples both Zimbabwean and British in support of her postcolonial argument, to the 'incongruous but arresting' symbols of chiefly office, and the equally incongruous role mixture of the modern, abode- and dress-code-swapping rural–urban commuter student/ chief. School teaching, like nursing, is one of the most enduring colonial traditions among Africans, in the sense of being one of the few skilled modern vocational/professional avenues open to African women. In this sense, apart from the fashionable clothes alluded to, there is nothing post-colonial about her chosen career; however, her assumption of the chief-tainship, and her decision and ability to combine this with a career, is certainly 'postcolonial' and simultaneously 'post-traditional'.

Perhaps unsurprisingly, Chief Mabhena's appointment continued to generate local controversy for some time. Her wedding in early 1998 gained much attention, including a one-page article in the London-based monthly magazine *New African*. Under a colour photograph of a very 'Western' modern wedding, which could have taken place anywhere in Africa, the article begins by stating that 'dressed in a stunning white gown, [she] tied the knot with Regiment Sibanda, a fellow school teacher, after a two-year romance'. It then summarises the controversy sur-rounding her appointment, with more pithy quotations from Welshman Mabhena ('This is an insult to our culture. We cannot allow a woman to become a chief') and others. She is supported by local women's groups, the ruling ZANU party's Women's League and President Mugabe. Ap-parently, she is now globally famous since giving a CNN TV interview, and as a result, a fan club in the USA 'is collecting money to buy her a car so she can use it to travel from her village to the school where she teaches' (*New African* 1998: 28). The pastiche of imagery and symbolism, the apparently contradictory but syncretic modes of articulation and posi-tioning, and her global renown through TV and print media coverage, again highlight aspects of postmodernism and postcolonialism but em-braced by the notion of post-traditionalism as explained above.

The implications of this perspective need to be considered, both in terms of terminology and in relation to frames of reference, not least by postcolonial authors concerned primarily with the identities and world views of former colonial subject peoples now living in erstwhile imperial metropolises in Europe and North America. From the examples just ana-lysed, and from *a priori* argument, it would seem to have several potential merits. First, it encompasses a broader, less specific sense of the 'tradi-tional' as the accumulated amalgam of practices and beliefs from pre-vious epochs and domains. Second, and following on directly from the previous point, it removes the 'colonial' nomenclature, which imposes an implicitly Northern-centric fixety of epoch and dominant peoples, identities, institutions and discourses/practices; third, it is historically more inclu-sive, since not all territories and indigenous polities in the South were col-onised by Europe, although the vast majority certainly were; and, fourth, it implies the possibility of greater weight being given to indigenous and hybrid pasts, which may in turn (re)combine in new hybrid ways, both appropriate and dysfunctional/disjunctural. Naturally, this perspective has much in common with some strands of postcolonial writing, which are concerned with the recovery of lost indigenous histories and identities. However, to the extent that it might signpost the importance of Southern perspectives and modes of problematisation and decentre Northern preoccupations, 'post-traditionalism' seems a helpful concept.

However, even this term is already being used in different contexts, which suggests the need for some caution (Simon 1998). It is also apposite in this context to consider the importance of shedding the colonial and/or 'traditional' modernist legacies within academic disciplines like geography, in the South, as we move towards the new millennium (Crush 1993; Simon 1994; Singh 1995).

Rethinking development: from theory to practice

The conceptualisations of postmodernism, postcolonialism and post-traditionalism that I have advanced are, of course, very different from the paralysing and conservatising, even self-indulgent, nature of extreme relativism adopted by some postmodernists in the North. I am trying to push out the frontiers of theorising and engagement. Therefore, in this concluding section I shall examine two questions, namely the extent to which these poststructuralist and postmodern ideas, and also current conceptualisations of sustainable development, inform how and what we teach our students, and, second, how far these ideas are being transferred into the arena of practice by the broader development community.

Teaching development studies

The volume of research publications attempting to engage with poststructural and postmodern perspectives is rising but still represents only a small fraction of total research output within development studies. As these paradigms become more widely accessible, one might expect student interest in development studies to wane somewhat. Perhaps paradoxically, the opposite seems true. Student numbers on undergraduate courses are holding firm, at least in the UK, the Netherlands and Nordic countries, while new postgraduate taught courses have been or are being established by a number of universities (see also Cline-Cole, Chapter 5 this volume). These are popular among British and foreign students alike.

Less surprising, then, is the current rush by publishers to produce new textbooks in development studies that attempt to keep up with the rapid recent changes in the South. An indication of how much has changed in a decade is that Leeson and Minogue's *Perspectives on Development*, arguably the leading development text when published in 1988, now feels somewhat dated, as it made no reference to the development impasse and predated the post-development debates. Its central aim of comparing the different perspectives of various social scientific disciplines on the theory and practice of development, is also less pertinent today than when it was published on account of greater cross-disciplinary work since then.

New volumes are appearing all the time, and the current choice is unprecedented. I will concentrate on those books which include some conceptual or theoretical coverage. In some cases, authors have sought to update older books through new editions. Such efforts are often unexciting, if only because they tend to retain the old format and approach (albeit within a redesigned jacket), merely adding a postscript chapter, updating some data or case studies, and rearranging some parts of the text at a time when a fresh approach is really called for. Perhaps the most relevant recent exemplar of this category is the second edition of Björn Hettne's *Development Theory and the Three Worlds*, which appeared in 1995. The publisher's blurb on the cover implies far greater revisions than the minor ones actually undertaken. Most conspicuous, moreover, is the paucity of new literature added to the bibliography and the absence of any mention of, let alone engagement with, poststructural and postmodern perspectives. Of the more empirically focused but conceptually informed books, the second edition of *A Geography of the Third World* by Dickenson *et al.* (1996) illustrates just what can be achieved by substantial rewriting and reorganisation.

Among the plethora of recent books on development, Paul Streeten (1995) and Marshall Wolfe (1996) have written retrospective accounts from the perspective of their own careers, highlighting the challenges of changing times and paradigms. Both express regret that more has not been achieved in the field of poverty alleviation but also point out some of the contemporary challenges, especially the necessity of having states that *are* capable of decisive action and development intervention when appropriate. In other words, they decry the current obsession with minimising the interventionist role of the state. Under the circumstances, it is not very surprising that they do not discuss poststructural perspectives. While not a valedictory survey prior to retirement, Colin Leys's *The Rise and Fall of Development Theory* (1996) has some similarities with the former two books, in as much as it is a collection of papers and chapters in the broad subject area, written at various stages over a number of years, although several are original to the book. There is much excellent material in an accessible prose style suitable for both teaching and research, but Leys does not address the recent theoretical 'turns'. The same is true of Ozay Mehmet (1995), whose slim volume is a rather disappointing critique of Eurocentrism in economic development theories but ultimately addresses only a few of the mainstream contenders. Arturo Escobar's *Encountering Development* (1995), one of the most widely distributed and read of the recent texts, is written as a detailed and elegant critique of conventional development(alism) from an anti-development position. However, his devastating critique is not matched by an exposition of an alternative vision or its application; even his final chapter – something of a prospect – does little more than eulogise NGOs as representing the way forward.

At least four recent textbooks on development theory, namely Peter Preston's *Development Theory: An Introduction* (1996), John Brohman's *Popular Development* (1996), John Martinussen's *Society, State and Market* (1997) and John Rapley's *Understanding Development* (1996), offer clearly written and accessible accounts of the various modern development theories and debates, linked to issues surrounding the role of the state and the impact of external and internal imperatives. However, they all omit any explicit reference to poststructuralist and postmodern theories. Rapley's final paragraph begs the question without even a hint of such possibilities:

> The time for hard questions is approaching. If the experience of the East Asian NICs was exceptional, if conditions have changed for the worse in the world economy and the international political economy, if the political and economic prospects for some countries are growing bleaker all the time, a serious reconsideration of what development is and should entail may be in order. The time for another paradigm shift may be drawing near.
>
> *(ibid.: 158)*

The development studies reader edited by Stuart Corbridge (1995) also omits any reference to poststructural debates. I am aware of only two substantial and four partial exceptions to this state of affairs. *Rethinking Social Development*, edited by David Booth (1994) and *Power of Development*, edited by Jonathan Crush (1995) offer the most sustained and accessible treatments of these perspectives and themes; the former is explicitly concerned mainly with methodological aspects of post-impasse development. As edited volumes, though, the extent of engagement and the precise perspective varies between chapters. Similarly, the first of my four partial exceptions, *The Diversity of Development* (Van Naerssen *et al.* 1997), contains several theoretical chapters (including one by Frans Schuurman) that do address current trends and debates. This volume is actually the *festschrift* for Jan Kleinpenning, a leading Dutch development geographer. John Brohman's (1996) text goes rather further than the other single- or double-authored texts in opening such vistas and considering 'alternative' issues and agendas which reject Eurocentric bias and grand theories, and centre on empowerment of the poor. Although he uses the umbrella term 'popular development', much of his analysis corresponds to post-developmentalism as discussed above. In a very recent contribution, Gilbert Rist (1997) has provided one of the most accessible and at the same time challenging accounts of the rise of development and developmentalism. However, it offers a differently structured history, not only emphasising the role of President Truman and the 'invention of development' but also devoting greater attention to alternatives within what is sometimes known as 'third worldism'. Rist addresses what he terms 'the postmodern illusion' in the penultimate chapter. While useful up to a point, this is limited by his unidimensional interpretation of postmodernism as consisting largely of the 'virtual reality' spawned by the telecommunications revolution and globalisation; however, he does link this to the genesis of ideas for moving beyond development. A distinctive feature of the book is its partial decentring of the Second World War and its immediate aftermath as the genesis of 'development'. As mentioned earlier, Rist draws attention to the importance of two earlier watersheds linked to the League of Nations, i.e. the mandates system and multilateral technical assistance to China. Finally, Michael Cowen and Robert Shenton (1996) seek to address postmodernism by way of a concluding chapter in their *Doc-*

trines of Development, which was clearly written as something of an after-thought and does not sit easily with the rest of their material. It certainly fails to engage with the new perspective, and is little more than heated invective against one particular edited volume, which Cowen and Shenton take to be the embodiment of all the evil that is postmodernism.[13]

It seems appropriate to include in this discussion two slightly older books, which, while not incorporating poststructural and postmodern theories explicitly, do set out detailed alternative, locally appropriate and bottom-up approaches to development, which they call empowerment and co-evolution respectively. These are John Friedmann's *Empowerment* (1992) – on which Brohman (1996) drew significantly, and Richard Norgaard's *Development Betrayed* (1994). So perhaps paradoxically, these two represent the most substantive treatments of alternative, poststructural 'revisionings', to borrow Norgaard's term.

Undoubtedly, I have inadvertently omitted other books, but on the basis of those cited here, it seems clear that there is as yet no comprehensive textbook coverage of the major emerging theoretical ferment in terms far broader than the impact of the debt crisis and policies of liberalisation and democratisation for communities on the ground and those concerned with improving their lot. The reasons for such omissions are no doubt varied, but they include a lack of personal familiarity on the part of the authors, their perception that postmodernism in particular is irrelevant to the South, and the undoubted difficulties involved in trying to synthesise extremely diverse and even contradictory literatures in a manner accessible to undergraduates. For the time being, then, our students will be reliant on more fragmented research-oriented literature if they are to gain an understanding of the 'post-' paradigms.

As pointed out earlier, the environment is one of the leading development topics of the 1980s and 1990s. One would therefore expect to find extensive coverage of environment–development issues in this new crop of texts, taking full account of the sustainability debate and the range of environmentalist theories or ideological positions, such as ecodevelopment, deep ecology, political ecology and the like (*cf.* contributions to Sachs 1993). However, I must confess to some astonishment at how marginal this theme is in most of the books under discussion. Frans Schuurman's (1993) book contains a very useful overview chapter on 'Sustainable development and the greening of development theory' by Bill Adams. He (Adams) also contributed a chapter on 'Green development theory?' to Jonathan Crush's (1995) edited collection, alongside a chapter by Gavin Williams on 'Modernizing Malthus: the World Bank, population control and the African environment'; the environment, environmentalism and sustainable development are also touched on elsewhere and are extensively referenced in the index. However, although Booth's (1994) collection has three chapters on rural development issues, there is no detailed treatment of the environment and sustainability; these are merely referred to briefly by several authors. The editor acknowledges this but

---

[13.] For more detailed reviews of some of these recent books, see Brown (1996) and Simon (1997).

points out that the book is not designed as a survey of contemporary themes and issues in development studies (see above).

In terms of single- or double-authored books, the situation is far more worrying. As discussed above, Escobar (1995) provides roughly twenty pages on sustainable development – a critique of the Brundtland Commission report and a summary of some alternative perspectives. Martinussen (1997) devotes one eighteen-page chapter out of twenty-five to 'Development with limited natural resources'; Brohman (1996) one chapter out of eleven on 'Environment and sustainability'; and Rist (1997) one chapter out of thirteen on 'The environment, or the new nature of 'development'.' Streeten (1995) provides similar coverage as part of his chapter on 'Global institutions for an interdependent world', as does Marshall Wolfe (1996), with a chapter entitled 'The environment enters the political arena'. Corbridge's (1995) reader also has only a single reading (out of twenty-seven) on environment–development issues – excerpts from a paper on Amazonia by Susanna Hecht – and the editor's discussion of it in his introduction to the relevant section does not link it to wider environmental debates. By contrast, Hettne (1995), Mehmet (1995), Preston (1996) and Rapley (1996) all provide only outline sketches in three or four pages of the environment–development interface and issues of sustainability, with only one or two index entries to match. Cowen and Shenton (1996) mention them only in passing, while the words 'environment' and 'sustainable development' do not even appear in the index to Leys (1996); there is also no coverage of these issues in the two or three chapters dealing with the important theoretical development debates in Kenya and 'Development theory and the African tragedy'.

While one might understand the omission of postmodernism and similar paradigms from most of the new texts for the reasons mentioned above, the generally modest or even negligible attention to environmental issues in the single- and double-authored development volumes must be a matter for grave concern. What coverage there is tends to be compartmentalised in a separate chapter, which, while arguably highlighting its importance, also fails to engage with the extent to which environmental concerns and theorisations have been integrated within or have engaged other development theories and paradigms. The wide influence of Blaikie (1985), Blaikie and Brookfield (1987) and Redclift (1987), for example, in establishing what is now widely known as political ecology (Bryant 1992, 1997), makes this silence all the more remarkable.

One very commendable recent attempt to fill this void is *Environment and the Developing World* by Avijit Gupta and Mukul Asher (1998). Unlike the other texts considered here, this one focuses entirely on the environment, highlighting the interface between environmental processes and the impacts of human interventions in contexts ranging from coastal zones to urban areas. Population and demographic issues are discussed, as are water resource and energy development, environmental economics and impact assessment, and global issues such as climate change and ozone depletion. Importantly, the history of global environmental concerns and international initiatives to address them, from the early part of this century to post-Rio developments, is explored over several chapters.

Notwithstanding this last-mentioned and overtly environmental approach to development, there remains an urgent need for more up-to-date development studies texts. The hype from publishers surrounding the value of the plethora of recent books is clearly exaggerated, whatever the particular merits of the individual volumes covered here. In this context, it is noteworthy that none of the books reviewed is concerned directly with the more practical issues of methods and techniques for undertaking development research, policy and project or programme implementation. This underscores the value of the few recent publications which are designed explicitly as research resources and manuals. Britha Mikkelsen's (1995) volume was one of the first to disseminate methods and techniques appropriate to the changing perspectives in development thinking. Topics she covers include problem analysis, the importance of popular knowledge (given practical application through participatory rural appraisal (PRA) methods, which receive a critical exposition), gender analysis, and ethics and interventions. More recently, Robert Chambers, who is generally credited with developing and popularising PRA, has published his own guide to that methodology (Chambers 1997) as a follow-up to his earlier and provocative *Challenging the Professions* (1993), which articulated the need for participatory approaches. The latest addition to the literature is an accessible edited collection by Alan Thomas *et al.* (1998). This Open University course text embraces a wide range of approaches and methods in a student-friendly format; as with the others just mentioned, it is likely to be appreciated by development students and practitioners alike. It is perhaps worth mentioning that Chambers' two books are but examples of a far more practically oriented set of publications produced by the Intermediate Technology Development Group.

Postmodern practice in development

One response to the crisis of representation and the challenge of postmodern practice has been to withdraw from field research, advocacy and development work, a step that, as I argued above, may appear to let those individuals feel vindicated but does not ultimately let them off the moral hook. It might also address one of the key concerns of Cowen and Shenton (1996) by re-establishing a clear distinction between a secular process of development over time and the explicit intention to develop. By contrast, nearly a decade ago, as the theoretical impasse became more widely debated, Michael Edwards (1989) fomented lively exchanges by questioning the relevance of development studies to what happens on the ground. He bemoaned the lack of impact of theories and vast research efforts on conditions on the ground. Four years later, he felt the situation to be less dire and saw progress in integrating theory and practice through greater interaction between the respective groups of actors, the development of new research methodologies such as the actor-oriented approach and techniques including participatory rural appraisal:

In the 'postmodern void' which faces development studies today, it is important to have some convincing theory to act as a counterweight to conventional economics. ... Is there in fact any alternative to an eclectic approach which examines everyday experience from a number of different points of view and then synthesises the results into higher-level explanation? ... So long as theory is constructed from real experience it will have explanatory power. But it will never be 'Grand theory' in the sense implied in the classical tradition. This may be a disappointment to academics, but is scarcely relevant to practitioners.

(Edwards 1993: 85; see also a revised version in Edwards 1994: 287)

He then concluded by re-emphasising two core ideas:

First, that the purpose of intellectual enquiry in this field of study is to promote the development of people denied access to knowledge, resources and power for hundreds of years. *Second*, that the most effective way of doing this is to unite understanding and action, or theory and practice, into a single process which puts people at the very centre of both. This is the real task for development theories in the 1990s.

(Edwards 1993: 90)

The revised version of this chapter ends slightly differently but conveys the same sentiments:

Do we really need more research, or just more action to implement what we already know? ... The central message of my original paper on 'Irrelevance' was that neither poor people, nor the organizations established to work with them, nor the situations in which they struggle to survive and prosper, should be treated as objects for examination by outsiders. Unless and until this point is accepted, academics and practitioners will not be able to develop an equal and supportive partnership. And if this enterprise continues to be frustrated we shall all, in the end, be the losers.

(Edwards 1994: 296)

This perspective conveys a clear sense of moral duty or commitment to engagement; in other words, development practitioners are intent on continuing their work as sensitively and responsibly as possible (see also Edwards 1994; Corbridge 1993b, 1994). To this end, appropriate theories that are grounded in the real world will be helpful; others will be discarded. In this context, Corbridge's conclusion regarding issues of postmodern concern for cultural relativism is worth repeating:

The post-modern dilemma is avoided as and when we accept that certain human needs and rights, at least, can be taken to be 'universal', and when we learn that in attending to these needs and rights we are not so much dictating to others as dictating to ourselves.

(Corbridge 1994: 112)

By means of the concrete examples in the previous section, I have sought to indicate the potential relevance of the new perspectives in a manner that is very different from the grand theorising, universalising and/or uncritical eulogising of NGOs and new social movements so evident in the work of many postmodern and postcolonial proponents. Such bodies may have many advantages and now operate within a far more favourable climate of reduced state involvement and increasing political tolerance. However, they represent no sinecure and have many drawbacks, including vulnerability to changing circumstances, the loss of key leaders and former external funding, and misconduct. The challenges facing the civic associations in South Africa during and following the transition away from apartheid illustrate all these dilemmas as they have to adapt from frontline resistance politics to more developmental and/or watchdog roles under a legitimate government. Development may have become a dirty word to some, as an intimate part of a particular geopolitical and economic 'project' that has wreaked much havoc. In that sense, perhaps, we need to transcend 'development', not by pulling down the shutters but by formulating different paths to the same end. It is certainly incumbent on us not to moralise about or to seek to represent others, distant or proximate, by the uncritical/unself-conscious projection of our own world views from a position of unequal power. Ultimately, this amounts to asking whether development work in practice retains any legitimacy and, if so, for and by whom?

A UN report published on 12 June 1997 suggests that the elimination of world poverty would cost a relatively modest £50 billion, less than the annual budget of the UK, for example (BBC Radio 4 *Today* programme 12/6/97; although not stated, this may refer to the 1997 *Human Development Report*). The report emphasises facilitation as the crucial mechanism – helping people to help themselves through appropriate, enabling interventions such as promoting education (especially of girls, who commonly have far lower enrolment, completion and literacy rates than boys) and debt relief rather than merely the untargeted giving of increased aid volumes. This approach has become widespread within progressive agencies and NGOs and is substantively different from the supposedly targeted aid conditionalities of structural adjustment and economic recovery programmes, where social expenditure cuts have retarded rather than accelerated skills acquisition and hence the development of civil society as a counterweight to government.

The implications here are clearly that *increased* responsive, co-operative, locally appropriate and directed resource transfers or partnerships do and can have a continuing role in meeting the challenge of widespread poverty. Alongside nutritional and other basic need considerations, powerlessness is a critical component of poverty. External engagement can play a key role in promoting empowerment, although this term too has now been popularised in development liturgy in often simplistic or naive ways which ignore the fact that households, villages and other communities are seldom homogeneous and that empowerment rarely occurs uniformly within them. What I am ultimately arguing is that it is the *basis of intervention* rather more than *whether* to intervene that is at stake. Naturally, this is *not* to favour some outdated missionary zeal imbued with Eurocentrism or naive unversalism, and directed at all and sundry, aiming to show 'them' the route to enlightenment or development heaven. However, we – the wealthy and powerful of both North and South – are the beneficiaries of the current (largely capitalist) world order, which produces poverty, misery and powerlessness alongside wealth and power in varying combinations in all parts of the globe. Therefore, for both material and ethical reasons, we cannot abandon our moral responsibility to the poor[14] unless we see ourselves as café patrons who studiously if somewhat uncomfortably ignore the pavement beggar or we regard the struggle for survival and development of the poor in (our own and) other cultures and countries as a leisuretime spectacle akin to a latter-day gladiatorial contest to be observed and discussed at a safe distance, albeit on television rather than in the Colosseum!

References

Adedeji, A. (ed.) (1993) *Africa Within the World: Beyond Dispossession and Dependence*. Zed Books, London.

Africa Confidential (1997) World apart, *Africa Confidential* **38**(1): 1–3.

Barff, R. (1995) Multinational corporations and the New International Division of Labor, in Johnston, R.J., Taylor, P.J. and Watts, M.J. (eds) *Geographies of Global Change: Remapping the World in the Late Twentieth Century*. Blackwell, Oxford.

Barratt Brown, M. (1997) An African road for development: are we all romantics? *Leeds African Studies Bulletin* No. 62. African Studies Unit, University of Leeds.

Bauer, P. (1972) *Dissent on Development*. Weidenfeld & Nicolson, London.

Bauman, Z. (1992) *Intimations of Postmodernity*. Routledge, London.

Bayart, J-F. (1993) *The State in Africa: The Politics of the Belly*. Longman, Harlow.

Bell, M. (1992) The water decade valedictory, New Delhi 1990: where pre- and post modernism met, *Area* **24**(1): 82–89.

Bell, M. (1994) Images, myths and alternative geographies of the Third World, in Gregory, D., Martin, R. and Smith, G. (eds) *Human Geography: Society, Space and Social Science*. Macmillan, London.

Berg, L.D. (1993) Between modernism and postmodernism, *Progress in Human Geography* **17**(4): 490–507.

Bhabha, H.K. (1994) *The Location of Culture*. Routledge, London.

Blaikie, P. (1985) *The Political Economy of Soil Erosion*. Longman, Harlow.

Blaikie, P. and **Brookfield, H.** (with others) (1987) *Land Degradation and Society*. Methuen, London.

[14.] Corbridge (1998) provides a fuller discussion of this issue, in the context of critiques by the New Right and the 'post-Left'.

Blaut, J.M. (1993) *The Colonizer's Model of the World: Geographical Diffusionism and Eurocentric History*. Guilford, New York.

Booth, D. (1985) Marxism and development sociology: interpreting the impasse, *World Development* **13**(7), 761–787.

Booth, D. (ed.) (1994) *Rethinking Social Development: Theory, Research and Practice*. Longman, Harlow.

Brohman, J. (1996) *Popular Development: Rethinking the Theory and Practice of Development*. Blackwell, London and Oxford.

Brown, E. (1996) Deconstructing development: alternative perspectives on the history of an idea, *Journal of Historical Geography* 22(3): 333–339.

Bryant, R. (1992) Political ecology: an emerging research agenda in Third World studies, *Political Geography* **11**(1): 12–36.

Bryant, R. (1997) Beyond the impasse: the power of political ecology in Third World environmental research, *Area* **29**(1): 5–19.

Carson, R. (1962) *Silent Spring*. Houghton Mifflin, New York.

Chambers, R. (1993) *Challenging the Professions: Frontiers for Rural Development*. Intermediate Technology Publications, London.

Chambers, R. (1997) *Whose Reality Counts? Putting the First Last*. Intermediate Technology Publications, London.

Colás, S. (1994) *Postmodernity in Latin America: The Argentine Paradigm*. Duke University Press, Durham, NC.

Corbridge, S. (1993a) Colonialism, post-colonialism and the political geography of the third world, in Taylor, P.J. (ed.) *Political Geography of the Twentieth Century*. Belhaven, London.

Corbridge, S. (1993b) Marxisms, modernities and moralities: development praxis and the claims of distant strangers, *Environment and Planning D: Society and Space* **11**: 449–472.

Corbridge, S. (1994) Post-Marxism and post-colonialism: the needs and rights of distant strangers, in Booth, D. (ed.) *Rethinking Social Development: Theory, Research and Practice*. Longman, Harlow.

Corbridge, S. (ed.) (1995) *Development Studies – A Reader*. Edward Arnold, London.

Corbridge, S. (1998) Development ethics: distance, difference, plausibility, *Ethics, Place and Environment* **1**(1): 35–53.

Cornia, G.A., Van der Hoeven, R. and **Mkandawira, T.** (eds) (1992) *Africa's Recovery in the 1990s: From Stagnation and Adjustment to Human Development*. Macmillan, London.

Cowen, M. and **Shenton, R.** (1996) *Doctrines of Development*. Routledge, London.

Crawford, L. (1996) Golf rebels seal off their town, *Financial Times*, 16–17 December. London.

Crush, J. (1993) The discomforts of distance: post-colonialism and South African geography, *South African Geographical Journal* **75**(2): 60–68.

Crush, J. (ed.) (1995) *Power of Development*. Routledge, London.

Culpeper, R. and **McAskie, C.** (eds) (1997) *Toward Autonomous Development in Africa*. North–South Institute, Ottawa.

Davidson, B. (1992) *The Black Man's Burden: The Curse of the African Nation-State*. James Currey, London.

Dear, M. (1988) The postmodern challenge, *Transactions of the Institute of British Geographers, New Series* **13**: 262–274.

Development Dialogue (1978) Special issue: *The Third World and Another Development*, 1978: 2.

Development Dialogue (1980) Special issue: *Another Development – Perspectives for the 'Eighties*, 1980: 1.

Dickenson, J.G. *et al.* (1996) *A Geography of the Third World* (2nd edn). Routledge, London.

Ecologist, The (1972) *A Blueprint for Survival*. Penguin, Harmondsworth.

Edwards, M. (1989) The irrelevance of development theory, *Third World Quarterly* **11**(1): 116–136.

Edwards, M. (1993) How relevant is development studies? in Schuurman, F. (ed.) *Beyond the Impasse: New Directions in Development Theory*. Zed Books, London.

Edwards, M. (1994) Rethinking social development: the search for relevance, in Booth, D. (ed.) *Rethinking Social Development: Theory, Research and Practice*. Longman, Harlow.

Edwards, M. and **Hulme, D.** (eds) (1995) *Non-Governmental Organisations: Performance and Accountability – Beyond the Magic Bullet*. Earthscan, London.

Ekins, P. (1992) *A New World Order: Grassroots Movements for Global Change*. Routledge, London.

Engberg-Pedersen, P., Gibbon, P., Raikes, P. and **Udsholt, L.** (eds) (1996) *Limits of Adjustment in Africa*. Centre for Development Research, Copenhagen, and James Currey, Oxford.

Escobar, A. (1995) *Encountering Development: The Making and Unmaking of the Third World*. Princeton University Press, Princeton, NJ.

Esteva, G. (1992) Development, in Sachs, W. (ed.) *The Development Dictionary*. Zed Books, London.

Featherstone, M. (1991) *Consumer Culture and Postmodernism*. Sage, London.

Featherstone, M. (1995) *Undoing Culture: Globalization, Postmodernism and Identity.* Sage, London.

Folch-Serra, M. (1989) Geography and post-modernism: linking humanism and development studies, *Canadian Geographer* **33**(1): 66–75.

Friedmann, J. (1992) *Empowerment.* Blackwell, Oxford.

Galli, R. (1992) Winners and losers in development and antidevelopment theory, in R. Galli (ed.) *Rethinking the Third World.* Crane Russak, New York.

Giddens, A. (1994) Beyond Left and Right: the Future of Radical Politics. Polity, London.

Gupta, A. and **Asher, M.G.** (1998) *Environment and the Developing World.* Wiley, Chichester.

Hallen, B. (1995) Some observations about philosophy, postmodernism and art in contemporary African studies, *African Studies Review* **38**(1): 69–80.

Hanlon, J. (1991) *Mozambique: Who Calls the Shots?* James Currey, London.

Hanlon, J. (1996) *Peace Without Profit: How the IMF Blocks Rebuilding in Mozambique.* James Currey, Oxford.

Harvey, D. (1989) *The Condition of Postmodernity.* Blackwell, Oxford.

Hettne, B. (1995) *Development Theory and the Three Worlds* (2nd edn). Longman, Harlow.

Himmelstrand, U., Kinyanjui, K. and **Mburugu, E.** (eds) (1994): *African Perspectives on Development.* James Currey, London.

Hirsch, F. (1977) *The Social Limits to Growth.* Routledge & Kegan Paul, London.

Hudock, A. (1995) Sustaining southern NGOs in resource-dependent environments, *Journal of International Development* **7**(4): 653–667.

Hurrell, A. (1995) International political theory and the global environment, in Booth, K. and Smith, S. (eds) *International Relations Theory Today.* Polity, London.

Husain, I. and **Faruqee, R.** (eds) (1996) *Adjustment in Africa: Lessons from Country Case Studies.* Avebury, Aldershot, for the World Bank.

Independent, The (1997) 28 June. London.

Lal, D. (1983) *The Poverty of 'Development Economics'.* Hobart Paperback 16, Institute of Economic Affairs, London.

Leeson, P.F. and **Minogue, M.M.** (eds) (1988) *Perspectives on Development: Cross-Disciplinary Themes in Development.* Manchester University Press, Manchester.

Lehmann, D. (1997) An opportunity lost: Escobar's deconstruction of development, *Journal of Development Studies* **33**(4): 568–578.

Leontidou, L. (1993) Postmodernism and the city: Mediterranean versions, *Urban Studies* **30**(6): 949–965.

Leys, C. (1996) *The Rise and Fall of Development Theory*. James Currey, Oxford.

Marchand, M. and **Parpart, J.** (eds) (1995) *Feminism/Postmodernism/Development*. Routledge, London.

Martinussen, J. (1997) *Society, State and Market: A Guide to Competing Theories of Development*. Zed Books, London.

McFadden, P. (1997) Gender equity and African development, in Culpeper, R. and McAskie, C. (eds) *Toward Autonomous Development in Africa*. North–South Institute, Ottawa.

McGee, T.G. (1997) The problem of identifying elephants, globalizations, and the multiplicities of development, Paper presented at the Lectures in Human Geography Series, Dept of Geography, University of St Andrews, Scotland, 30–31 May.

Meadows, D.H., Meadows, D.L., Randers, J. and **Behrens, W.** (1972) *The Limits to Growth*. Pan, London.

Mehmet, O. (1995) *Westernizing the Third World: The Eurocentricity of Economic Development Theories*. Routledge, London.

Meldrum, A. (1996) The chief who wears a miniskirt, *The Guardian*, 24 December. London.

Middleton, N., O'Keefe, P. and **Moyo, S.** (1993) *Tears of the Crocodile: From Rio to Reality in the Developing World*. Pluto, London.

Mikkelsen, B. (1995) *Methods for Development Work and Research: A Guide for Practitioners*. Sage, New Delhi.

Miles, M. and **Crush, J.** (1993) Personal narratives as interactive texts: collecting and interpreting migrant life-histories, *Professional Geographer* **45**(1): 84–94.

Minh-ha, T.T. (1989) *Woman, Native, Other: Writing Postcoloniality and Feminism*. Indiana University Press, Bloomington.

Mitchell, T. (1995) The object of development: America's Egypt, in Crush, J. (ed.) *Power of Development*. Routledge, London.

Moser, C.O.N. (1993) *Gender, Planning and Development: Theory, Practice and Training*. Routledge, London.

Nabudere, D.W. (1997) Beyond modernization and development, or why the poor reject development, *Geografiska Annaler* **79B**(4): 203–215.

Namibian, The (1997) 6 June. Windhoek.

Nederveen Pieterse, J. (1991) Dilemmas of development discourse: the crisis of developmentalism and the comparative method, *Development and Change* **22**(1): 5–29.

Nederveen Pieterse, J. (1998) My paradigm or yours? Alternative development, post-development, reflexive development, *Development and Change* **29**(2): 343–373.

Nesmith, C. and **Radcliffe, S.A.** (1993) (Re)mapping mother Earth: a geographical perspective on environmental feminisms, *Environment and Planning D: Society and Space* **11**: 379–394.

New African (1998) Chief Mabhena defies her critics, No. 362 (April): 28.

Norgaard, R.B. (1994) *Development Betrayed: The End of Progress and a Co-evolutionary revisioning of the future.* Routledge, London.

Parayil, G. (1996) The 'Kerala model' of development: development and sustainability in the Third World, *Third World Quarterly* **17**(5): 941–957.

Parnwell, M.J. (1998) Between theory and reality: the area specialist and the study of development, Chapter 4, this volume.

Penna, D.R. and **Campbell, P.J.** (1998) Human rights and culture: beyond universality and relativism, *Third World Quarterly* **19**(1): 7–27.

Porter, D.J. (1995) Scenes from childhood; the homesickness of development discourses, in Crush, J. (ed.) *Power of Development.* Routledge, London.

Preston, P.W. (1996) *Development Theory: An Introduction.* Blackwell, London and Oxford.

Radcliffe, S.A. and **Westwood, S.** (eds) (1993) *Viva! Women and Popular Protest in Latin America.* Routledge, London.

Rapley, J. (1996) *Understanding Development: Theory and Practice in the Third World.* UCL Press, London.

Redclift, M. (1987) *Sustainable Development.* Methuen, London.

Reed, D. (ed.) (1992) *Structural Adjustment and the Environment.* Earthscan, London.

Rist, G. (1997) *The History of Development: From Western Origins to Global Faith.* Zed Books, London.

Routledge, P. (1993) *Terrains of Resistance: Nonviolent Social Movements and the Contestation of Place in India.* Praeger, Westport, Conn.

Routledge, P. (1995) Resisting and reshaping the modern: social movements and the development process, in Johnston, R.J., Taylor, P.J. and Watts, M.J. (eds) *Geographies of Global Change: Remapping the World in the Late Twentieth Century.* Blackwell, Oxford.

Sachs, W. (ed.) (1992) *The Development Dictionary: A Guide to Knowledge as Power.* Zed Books, London.

Sachs, W. (ed.) (1993) *Global Ecology: A New Arena of Political Conflict.* Zed Books, London.

Schumacher, E.F. (1973) *Small is Beautiful: A Study of Economics as if People Mattered.* Abacus, London.

Schuurman, F. (1993a) Modernity, post-modernity and the new social movements, in Schuurman, F. (ed.) *Beyond the Impasse: New Directions in Development Theory.* Zed Books, London.

Schuurman, F. (ed.) (1993b) *Beyond the Impasse: New Directions in Development Theory.* Zed Books, London.

Schuurman, F. and **Van Naerssen, T.** (eds) (1989) *Urban Social Movements in the Third World.* Routledge, London.

Shiva, V. (1988) *Staying Alive: Women, Ecology and Development.* Zed Books, London.

Sidaway, J. and **Power, M.** (1995) Sociospatial transformations in the 'postsocialist' periphery: the case of Maputo, Mozambique, *Environment and Planning A* **27**: 1463–1491.

Simon, D. (1989) Sustainable development: theoretical construct or attainable goal? *Environmental Conservation* **16**(1): 41–48.

Simon, D. (1994) Putting South Africa(n geography) back into Africa, *Area* **26**(3): 296–300.

Simon, D. (1995a) Debt, democracy and development: sub-Saharan Africa in the 1990s, in Simon, D., Van Spengen, W., Dixon, C. and Närman, A. (eds) *Structurally Adjusted Africa: Poverty, Debt and Basic Needs.* Pluto, London.

Simon, D. (1995b) The medicine man cometh: diagnoses and prescriptions for Africa's ills, *Third World Quarterly* **16**(2): 319–325.

Simon, D. (1997) Unlocking Third World potentials, *Times Higher Education Supplement*, 24 January: 26.

Simon, D. (1998) Rethinking (post)modernism, postcolonialism and post-traditionalism: South–North perspectives, *Environment and Planning D: Society and Space* **16**(2): 219–245.

Simon, D., Van Spengen, W., Dixon, C. and **Närman, A.** (eds) (1995) *Structurally Adjusted Africa: Poverty, Debt and Basic Needs.* Pluto, London.

Singh, R.P.B. (1995) Identity of geography in India beyond 2000: a search of sensitivity for future, *National Geographical Journal of India* **41**(3): 263–281.

Slater, D. (1992a) Theories of development and politics of the postmodern – exploring a border zone, *Development and Change* **23**(3): 283–319.

Slater, D. (1992b) On the borders of social theory: learning from other regions, *Environment and Planning D: Society and Space* **10**(3): 307–327.

Slater, D. (1993) The geopolitical imagination and the enframing of development theory, *Transactions of the Institute of British Geographers, New Series* **18**(4): 419–437.

Slater, D. (1995a) Challenging western visions of the global: the geopolitics of theory and North–South relations, *European Journal of Development Research* **7**(2): 366–388.

Slater, D. (1995b) Trajectories of development theory: capitalism, socialism, and beyond, in Johnston, R.J., Taylor, P.J. and Watts, M.J. (eds) *Geographies of Global Change: Remapping the World in the Late Twentieth Century*. Blackwell, Oxford.

Slater, D. (1997) Spatialities of power and postmodern ethics – rethinking geopolitical encounters, *Environment and Planning D: Society and Space* **15**(1): 55–72.

Soja, E. (1991) *Postmodern Geographies*. Verso, London.

Soja, E. (1996) *Thirdspace: Journeys to Los Angeles and Other Real-and-Imagined Places*. Blackwell, Oxford.

Streeten, P. (1995) *Thinking about Development*. Cambridge University Press, Cambridge.

Tapscott, C. (1995) Changing discourses of development in South Africa, in Crush, J. (ed.) *Power of Development*. Routledge, London.

Taylor, P.J. (1996) *The Way the Modern World Works: World Hegemony to World Impasse*. Wiley, Chichester.

Thomas, A., Chataway, J. and **Wuyts, M.** (eds) (1998) *Finding out Fast: Investigative Skills for Policy and Development*. Sage, London.

Thrift, N. (1995) A hyperactive world, in Johnston, R.J., Taylor, P.J. and Watts, M.J. (eds) *Geographies of Global Change: Remapping the World in the Late Twentieth Century*. Blackwell, Oxford.

Townsend, J.G. (1995) *Women's Voices from the Rainforest*. Routledge, London.

Trainer, F.E. (1996) Why development is impossible, *Scandinavian Journal of Development Alternatives and Area Studies* **15**(3&4): 73–89.

United Nations Development Programme (annual) *Human Development Report*. Oxford University Press, New York.

Van Naerssen, T., Rutten, M. and **Zoomers, A.** (eds) (1997) *The Diversity of Development*. Van Gorcum, Assen.

Walker, R.B.J. (1988) *One World, Many Worlds: Struggles for a Just World Peace*. Rienner, Boulder, Colo.

Ward, B. and **Dubos, R.** (1972) *Only One Earth: The Care and Maintenance of a Small Planet*. Penguin, Harmondsworth.

Watson, S. and **Gibson, K.** (eds) (1995) *Postmodern Cities and Spaces*. Blackwell, Oxford.

Whatmore, S. (1995) From farming to agribusiness: the global agro-food system, in Johnston, R.J., Taylor, P.J. and Watts, M.J. (eds) *Geographies of Global Change: Remapping the World in the Late Twentieth Century*. Blackwell, Oxford.

Wisner, B. (1988) *Power and Need in Africa*. Earthscan, London.

Wolfe, M. (1996) *Elusive Development*. Zed Books, London.

Woodward, D. (1992) *Debt, Adjustment and Poverty in Developing Countries* (2 vols). Belhaven, London.

World Bank (annual) *World Development Report*. Oxford University Press, New York.

World Commission on Environment and Development (1987) *Our Common Future*. Oxford University Press, New York.

CHAPTER 3

Rethinking development in India
Perspective, crisis and prospects

Rana P.B. Singh

Introduction

In spite of tremendous scientific–technological innovations and their application in the name of development, humankind today faces a crisis at several levels. The main idea of 'progress and development' has been conceived in the light of the positivistic–reductionist, empiricist and anthropocentric enterprise introduced in Europe in the seventeenth century. However, the alternative world view of interconnectedness and holism is considered more relevant today. Modern science, like new physics, is recapitulating in many ways the ancient spiritual world view of integral living. In the light of such shifts, the idea of 'development' also needs to be reassessed, rethought and reoriented.

The sacred world view is a unitary principle of cosmic interconnectedness in search of universal order and meaning of existence – ecological cosmology. Is it possible to discern any sense of sacred in the 'alternative development' strategy? Eco-friendly development is a strategy that emphasises the integration and promotion of environmental sustainability by reorienting lifestyles, transforming technology to suit the socio-economic and indigenous systems of resource use and local habitat, maximising employment opportunities through a structural shift in occupational structure, and stimulating a 'trickle down' process to ensure socio-spatial equity. The world is full of objects and ideologies which 'we do not notice, either because we are not interested in them or because we are not equipped to see them' (Paden 1992: 7). We take into account what currently concerns us, and avoid matters that have deeper values and that may have application in the distant future. Remember, the present makes the future and also interlinks with the past. Interpretation is sometimes connected with scientific viewpoints of causal explanation; nevertheless, it is also associated with the humanistic explanation of meanings where stands the sense of sacredness.

Like two sources of knowledge – the 'indirect' notion of understanding and imagination, and the direct one related to sensibility and perception – 'development' is also to be purveyed in the light of non-sensuous and sensuous sources, and these might be seen as complementary. This ideology has a presuppositional bias. Each culture has some of its own distinctiveness but also commonality with others. One cannot presume that any one culture is complete and to be adopted by or superimposed upon others. Certainly, we need a thorough diagnosis of the sick culture, instead of always prescribing and using different means, measures and medicines for recovery in the frame of 'trial and error'. Let us understand the total spectrum of disordering.

In the process of development since independence in 1947, we have encouraged cultures of consumerism, inequality, exploitation of natural resources, falling human values, immorality, social injustice and insecurity, loss of mutual cohesiveness, individualism and self-centredness, materialism, violence, environmental pollution, loss of tribal and traditional lifestyles, unemployment, and so on. In a recent television programme (on the DD1 National channel, 12 February 1998 at 10.30 pm), the leader of an environmental awareness movement, Ms Medha Patekar, argued for national co-ordination between public movements with an activist political consciousness and made an appeal for all political parties to replace their fostering of and fawning upon a political leader with a spiritual concept of human development and national integrity. Generally, conventional development is occurring only within the purview of economics, and we have reached the point where the freedom of society and our economic independence are in danger. Rural artisans, tribal culture, women and poor labourers have lost their status and roles in running the societal base and maintaining communal harmony. Natural and environmental settings have been ruined by the encroachment of the modern pace of development programmes like big dams, heavy industries and urban sprawl. The victim families are now living in slums and refugee camps or are wandering from one place to another in the hope of rehabilitation and a better life.

I do not want to write as an expert or scholar, but as a citizen of my culture and era, i.e. what India is at present, what roots it sustains and how it is trying to make its future. This is a narration of the intense 'inside' story – the never-ending political story – which I shall explain by an analogy based on the great epic, the *Ramayana*:

> On the orders of (king) Lord Rama, his brother Lakshmana paid a visit to the demon Ravana who was mortally wounded by Rama and was passing the last breath of his life. Ravana was the greatest scholar of the science of diplomacy and government. Lakshmana begged Ravana to teach him some of the key lessons of handling diplomacy and the smooth running of government.
>
> Ravana suggested, 'whatever system of government – democratic or dictatorship – is imposed is immaterial; but how you expose it, is a matter of serious concern. Make the public fool in such a way that the public are always confused. And whenever some groups raise their voices against it, convince them that you are trying to integrate the varying ideologies together; and form a Commission to formulate this plan. Continue this at least for five consecutive years. Afterwards take

time for reviewing and assessing progress; again make further plans and propagate assurances for the next five years. Take care that the public in no way learns which form of government is operating. This way you will succeed to rule this great country of India for ever. Remember that this culture has a deep sense of faith in the future, hope and tolerance! Take heed of this sense and rule them.'

Soon afterwards the demon, Ravana, died. Since then Rama is the president and Lakshmana is the prime minister of India in different faces and in different dresses.

Towards development and rethinking

The concept of 'development' has been based on the role of science and technology in the service of humankind with a view to controlling nature and the cosmic order. This needs reorientation – a paradigm shift. The political scientist Kothari (1990: 34) suggests: 'This must give place to the original purpose of science, namely seeking to understand the mysteries of nature with a deep sense of humanity and wonder.' However, it does not mean that humanity must sacrifice all the structures developed in the process of evolution and development. There is a need for *re*-evaluation and *re*orientation in the light of historical roots wherein lies the moral sense of development. Let us make general consciousness and awareness interlink with the concept of sanctifying the Earth and with community development. If there should be a moral imperative for sustainable development, we need a sense of sanctity about the Earth (*ibid.*: 33).

*Re*thinking development involves a shift from the present-day development crisis to the ethical march. A new ethic – value and faith systems – is required for humankind to live in harmony with the natural world on which we depend for survival and well-being. The rethinking should be based on the foundational value – the reasoning that underlies the ethical sense. Important ethical systems are based essentially on a foundational value, which for Gandhi was *ahimsa* (non-violence, which he interpreted as 'firmness in the truth'), for Schweitzer reverence for life, and for Aldo Leopold the sacredness of land (*cf.* Skolimowski 1990: 98).

The Gandhian view of non-violence, vegetarianism and *karma* (right action) is based on the idea of the total spiritual interconnectedness of all life, of the divine within all life. The essence of divinity exists in all life forms; all these should therefore be treated in the light of harmonious community life. Divinity is a deep sense of realisation of the power of interconnectedness with nature. All natural phenomena are, therefore, divine, sacred and of equal value. As human beings, we have to take the main responsibilities towards nature through a moral–ethical–religious approach. Gandhian thought is more relevant today and well suited to the soul and society of India. It is only possible if the controllers and legislators of development strategies are conscious of 'Self-realisation' (*svachetana*). This is different from the Western concept of 'ego-realisation'. The concept of 'Self-realisation' is known in the history of philosophy under various names, like 'the universal self', 'the absolute', 'the *atman*', etc.;

however, it is not 'self-centred'. This concept is the assumption of a capacity for self-determination, a capacity for realising potentialities (see Naess 1989: 85, 141).

One has to realise that 'the living Earth has a right to life, and that right is the primary moral argument for sustainable life' (Kothari 1990: 32). We need to make development sustainable, both environmentally and culturally. In this way, development also has a dimension of faith and reverence. However, ethical and moral pollution by materialism and consumerism is replacing the old value systems, which supported sustainability. Alas, this tendency of pollution is constantly threatening the sense of culture and even the identity of humankind (Singh 1996: 103).

In the Indian psyche, there are two image-views of society: (1) the very rich, who are considered to be like earthly gods as a result of their good *karma* in the past, and are perceived as happy people by others, and (2) the rest, who are poor, always trying to experience some pleasure through monetary gain. The first group donate a lot of money in the name of gods and to make their shelters (temples), while the second group worship there and always pray to the gods living there to be kind enough to bless them with wealth. Even the kingdom of God is controlled by rich and powerful humans. The power is mostly derived from money, with the further support of political power. 'Since money is everything, everything is to be done for money' is currently accepted as the social norm. Moreover, all the ethical norms are defined through the power of money. This is an absolute practice of materialism and its modern offshoot of consumerism. If someone slips down from the track, society labels it corruption and 'taking the wrong path'. However, remember that philosopher A.W. Watts (1962: 58) says: 'For in this world nothing is wrong, nothing is even stupid. The sense of wrong is simply failure to see where something fits into a pattern.'

The institutions involved in development at state or central level are reduced to the function of accounting for the allocation of funds between different government agencies, and they play only a small role in genuinely integrating the process of development (Gadgil and Guha 1996: 48). This issue is concerned more with political and personal interest, which is practised on priority level at the place of national development. 'Small is beautiful' is not always good; nor are 'big projects' always bad; rather these are a matter of context and regional suitability. It is a tyranny that in 'India the indigenous is sometimes the foreign, and the foreign is sometimes the indigenous' (Larson 1995: 144). We have left far behind the ethical domain of the Earth as the mother who nourishes us and whom we have to take care of; this is fully narrated in the *Prithvi Sukta* (Hymn to the Earth) of the *Atharva Veda*, a *c.* fifteenth century BCE text. Instead, in many respects we have accepted the Christian ideology of an exploitative and materialistic view of the Earth as resource. This is now part of our culture, and it has become difficult to escape from. Over time, by the process of development, the masses of Indian society have accepted corruption, inflation and rising prices, and social evils like the dowry and caste systems as part of their life and culture. Every Indian criticises them; nevertheless, we are part of it and are responsible for making them flourish.

The scenario and the roots

At some point in the present era, Indian culture lost its traditional sense of social harmony and community. *Dharma* (moral order, *not* religious dogma), which was the root of the culture, is now threatened by politically vested interests under the umbrella of secularism. Religion is currently used as a tool for 'secular democracy' with the support of 'secularisation of religious ideas'. The Indologist J. Larson (1995: x) opines that the post-independence 'secular state of India' is to a 'significant degree a forward-caste Neo-Hindu state, or, in other words, that the "secular state" in the Indian context has a number of religious aspects and may even represent in some respects a religious entity.' The philosopher S. Radhakrishnan (1956: vii–viii), who was the second president of India, argued that the Indian idea of secularism 'does not mean irreligion or atheism or even material comforts. It proclaims that it lays stress on universality of spiritual values which may be attained in a variety of ways. ... This is the meaning of a secular conception of the state though it is not generally understood.' In maintaining secular democracy, the Hindu sense of tolerance, a living tradition, 'has contributed vitally. ... More important, is the attitude of "live and let live" toward all manifestations of religious diversity' (Smith, 1993: 149). According to a recent newspaper report (*Aj*, 8 June 1997: 6), the central government provides a subsidy of Rs400 million (US$11.4 million) to Muslims for the annual Hajj (pilgrimage) to Mecca, and additionally it grants a sum of Rs930 million (US$26.6 million) for the arrangement of air and land transportation. For no other religious group are such provisions made. Would this be called a policy of 'religious secularism'? This is an example of satisfying a group (a so-called minority!) for electoral populism. At present the scene is different – more complex, together with crises and chaos. Of course, the majority of people still have a sense of hope for a miracle in the near future.

India has its own history of contrasts – ecological, religious, linguistic, historical, political and eco-psychological. At present, India is divided into thirty-three states, embracing 190 religious groups with 1,652 languages and dialects in twelve language families with twenty-four different scripts, and 3,742 castes and sub-castes further grouped into 4,635 communities. Before becoming a British colony, the country had never been a united sovereign state; instead, it had been an amalgam of various independent republics. Historically, successive invaders (e.g. the Huns, Turks, Kushans, Mughals) became part of this culture, and Hindu society accepted them all. It was only the British who rarely tried to be part of this culture and sought to remain aloof. The Hindu sense of tolerance was used as a tool to empower them, as it is the very material form from which the Indian character is formed (*cf.* Segal 1965: 33). The Hindu religion teaches a renunciation of worldly goods and preaches against materialism and consumerism. This spirit serves as a great source of strength in achieving sustainable development. However, it would be alarming if this spirit were to turn into fundamentalism (*cf.* Dwivedi 1990: 211).

When the British colonies in Asia and Africa became independent from the late 1940s onwards, they looked firmly away from their past and towards the future. India received independence from the British Empire on 15 August 1947. Since then, it has regarded the colonial past as best forgotten, a history of exploitation and humiliation that had left its people poor. Attention has been paid to economic development and prosperity, the benefits of which can supposedly be passed on to the poor masses. Strategies and plans were made to solve the then development problems through assessing the resource potential and spatial gaps. Following a 'socialist pattern of society' under the leadership of Jawaharlal Nehru, the first prime minister of the Indian republic, people from widely diverging interests and ideologies like Gandhians, Western-style socialists, communists, capitalist industrialists and others started working together and finally supported a central planning framework. This resulted in the formation of the Planning Commission of India, which introduced a series of Five-Year Plans, the first of which covered the period 1950–55. With a view to promoting central and state planning linkages, the National Development Council was set up in 1952. Currently, under the IXth Five-Year Plan (1997–2002) these development programmes are in progress. Until the Vth Plan (1974–79), more emphasis was laid on the national economy, eliminating poverty and promoting self-reliance, but in the following Plans the focus was changed with a view to benefiting the then government's policies and its romantic propaganda. A poet in the poets' assembly exclaimed about such plans (*Aj*, 13 June 1997: 3):

> There is everything in the new proposals
> To sell this country at the rate of salt.

The greatest loss recorded in India during the colonial period was the loss of our old ethic of eco-justice, which refers to the sanctity of life and cosmic interconnectedness (ecological cosmology), which extends to the sense of global family or universal brotherhood (*vasudhaiva kutumbakam*). This ethic helped society to maintain an order between *dharma* (moral code of conduct) and *karma* (right action). In the course of acculturation, the ideology of materialism, consumerism and individualism – which was always proscribed in traditional Hindu thought – has been accepted by contemporary society. At the other extreme – and perhaps in consequence – the movement of revival of ancient cultural values is being turned to fundamentalism by some groups. The old principle of *satyameva jayate* (only truth triumphs) is now replaced by *arthameva jayate* (only wealth triumphs). The foreign cultural domination of India during the last 700 years and the influence of imported culture have played a major role in this form of transformation (Dwivedi 1990: 210).

The country is currently facing a cultural crisis in most of its sectors. As a consequence of the progress of development, the rich are becoming richer and the poor poorer. It is a common complaint that whatever good ideas one has are highlighted in India's files of planning programmes, but the most crucial problem we face is non-implementation. The cultural and ethical crises that the country faces today are more dangerous than any other pollution. The good, age-old cultural traditions are not being followed, nor are the ills of the present tradition being solved. The culture of individualism, consumerism and sordid love for materialistic gain is part of our lifeworld. The political environment is facing the more critical scenario of cultural downfall, immorality and lack of national identity. Currently, the Indian political scene is structured by the scandalisation of the political system, the criminalisation of politics, the politicisation of criminals, and the amoralisation of political behaviour. In common talk, people say, 'Cash, Crime, Corruption, Communality and Casteism are the fundamental weapons for political victory in India' (*cf.* Singh and Singh 1981: 65). Segal's (1965: 309) advocacy is a warning: 'Will India have an Indian government ruthless enough to rule in the interest of the still-silent masses? And if it will not, how much longer will the masses stay silent?' How can this 'political environment' be cleaned? How is the problem of 'moral pollution' to be solved? Where has our philosophy of *karma* (right action) and *dharma* (right duty) gone? This is difficult to answer. Is it really a democracy that has to be regulated *of* the people, *by* the people and *for* the people? This is now replaced by the practice of 'off' the people, 'bye' the people and 'far' from the people .

Indian politicians operate in a manner similar to the Orwellian view of betrayal. Orwell's paraphrased statement (1945: 11) will help us to understand the situation:

> Never listen when they tell you that *politicians* and the *people* have a common interest, that the prosperity of the one is the prosperity of the others. It is all lies. A *politician* serves the interests of no creature except himself. And among us *people* let there be perfect unity, perfect comradeship in the struggle. All *politicians* are enemies. All people are comrades.

Politicians are successful in dividing the social psyche in terms of rich *vs* poor, owners *vs* labourers, high *vs* low castes, religious groupings, and so on. The policies of reservation and economic support for a section of society create conflict in the society due to discrepancies in their operation. In this process, politicians gain enhanced electoral support. There is at least hope for perfect unity among the various sects, sections and sectors of society. The policy of 'divide and rule' has taken root in the political environment. Larson (1995: 288) says:

> That India has made almost no progress in developing a uniform code and that it continues routinely to extend the system of 'reservations' even though such reservations were to cease ten years after independence (provides) further evidence of the politics of a nation without 'citizens'.

There is little hope that the polluted political environment will be cleaned and made more humane in the near future. Before any treatment of the sick culture and sick economy of India, a sympathetic diagnosis is necessary at all levels. During the late 1930s, the national poet, Maithilisaran Gupta, lamented: 'Let this educational system be dead as it is only for jobs.' Currently, there are 203 universities and 9,700 postgraduate colleges; however, this has been a great challenge for our education, which has left no place for ethical values and human service. Those who control power pontificate to others, while in their own life they practise something else. 'Corruption, like sacrifice, starts at the top and, percolating down, colours the whole society' (Segal 1995: 288). Bribery – or what is delicately called 'grease money' – is now so much rooted in the official functioning that without it nothing could move in the right direction. The rapid expansion of economic liberalisation and the commercialisation of politics are the crucial factors making corruption more intense.

Even after fifty years of independence, the common people in India feel that they are suppressed by the present 'New-Anglicised Indian rulers' (modern politicians and bureaucrats). Following the colonial policy of 'divide and rule', these rulers have also succeeded in segregating Indian society into various hierarchical sections of castes, cults, communities, religious sects and regions. Political culture is really, at present, a form of hypocrisy. In the name of Gandhian moralist thought, seminars and meetings are organised and declarations are made to follow the path shown by him. However, in practice nobody follows. False and 'dummy' celebrations are performed in which self-praise and big shows are the common scenes. In Orwellian thought, this is an example of 'doublethink'.

Some people say 'the British exploited India for the sake of their own country, but the present Indian rulers are exploiting us for individual gain; we are relieved from the British slavery, but in no way could we liberate ourselves from the "new" slavery imposed upon us by the political system after independence'. In British India, it was said that 'every Indian is born in debt and dies in debt'. In 1970, India's foreign debt was US$7,940 million, i.e. Rs110.5 per person; it reached US$106.7 billion in 1997, i.e. Rs4,210.5 (US$112.3) per person (Bose 1997: 52). The profit from such international debt accrues to some individuals and institutions, both at home and abroad, but every citizen has to bear a share of the burden. Budget deficits and escalating prices to consumers are the norm. Under the tough conditions of the IMF, India is bound to take loans, which are further misused by a particular section for their own profit. Ultimately, that burden of indebtedness falls on the common people.

In the name of development for depressed classes (*dalits*), the policy of reservation (22.5 per cent of all jobs) was introduced in the early 1950s through the initiatives of B.R. Ambedkar (1891–1956), the leader of the untouchables and India's first law minister. According to Article 334 of the Indian Constitution, this reservation was to be for ten years only. However, in 1960, 1970, 1980 and 1990 the time limit was routinely extended for an additional ten years, mostly to serve the vested interest of politicians for the sake of electoral support. In 1990, the Mandal Commission recommended an additional 27 per cent of reservations for OBCs (other backward castes). Thus, in total, 49.5 per cent of reservations have been approved by the central government. This policy and the idea of social justice are used as political tools. Comparing the statistics of employment by caste in 1935 and 1982, Khushwant Singh (1990: 19) concluded that 3.5 per cent of India's brahmin community (the highest caste in the hierarchy) holds between 36 and 63 per cent of all the plum jobs available in the country. Out of sixteen Indian prime ministers during the period 1947–98, eleven have been brahmins. Of course, under the reservation policies the people from the downtrodden classes are given opportunities even without the required qualification and merit. The upper-caste elites always complain about such an imposition. They suggest that the reservation policy should be applied only to education and scientific training, and that merit should be given priority in job selection. This is the way that, over time, the depressed classes should be upgraded. Through the misuse of reservation policy a great mass of unqualified engineers, doctors, administrators and even scientists (socalled!) are now part of the system by which the path of India's future development is to be made. One has to appreciate the level of tolerance of Indian people, who believe that this is the era of falsehood (*kali yuga*), where anything is possible. They say: wait for the new light; don't cry! This way the country is continuing on the path of development in the hope of miracle and charisma! There may be total collapse, or a completely new world!

According to the last census of India (1991), 74.3 per cent of the total 844 million inhabitants lived in rural areas, and 37.7 per cent represent the working population. The economy of the country is grouped into three sectors: (1) primary sector (farming, forestry, fishing, livestock and mining), (2) secondary (industry and manufacturing), and (3) tertiary (transport and communication, administration, and services). Their growth rates during 1950–1996 were 2.66, 5.6 and 5.1 per cent, respectively. Moreover, the primary sector's share in gross domestic product fell from 56.45 per cent in 1950–51 to 28.75 per cent in 1995–96. For the poor, lentils have been the major source of food protein; their share of daily food intake was 16.1 per cent in 1950–51, but this had declined to 7.1 per cent in 1995–96. Even at present, about 40 per cent of the people (mostly living in rural areas) do not have an adequate food intake (*cf.* Bose 1997: 29). In rural India, 34 per cent of families are still absolutely landless; with additional landless labourers, this reaches 39.5 per cent. When marginal landholders are included, the figure comes close to 58 per cent; together they possess less than 6 per cent of arable land (1992 data). Conversely, the top 5.3 per cent of landlords account for 40.17 per cent of total holdings. Even today, around 200 million people engaged in agriculture have no work for about 100 days per year (*ibid.*: 43). During the period 1981–91, the percentage of cultivators decreased from 41.58 to 38.75 per cent and the share of agricultural labourers increased from 24.94 to 26.15 per cent. It is sad to notice that land reforms over much of the country since independence have failed.

This has encouraged leftist movements, beginning with the communistled peasant revolts of Telangana in the 1940s and continuing at present with several armed organisations such as the Naxalites and similar extreme left movements among the tribal and poor landless farmers. The British land policy of high taxation, converting agricultural land into a commodity, creating a large class of tenant farmers, encouraging local chieftains to control the land and its products, and related factors promoted a substantial degree of autonomy and control by the landlords (*zamindars*). This resulted in the development of a two-ladder social class – the masters, omnivores, and the servants, dispossessed. The government policy of boosting agricultural production in limited areas like the northern states of Haryana and Punjab, with the support of conservative landlords, left the masses of people in the Indian countryside impoverished, while on the other side a new class of omnivore was created from among the large and medium-sized landholders of the areas of 'Green Revolution' agriculture (Gadgil and Guha 1995: 66). This lopsided agricultural development has had a series of environmental consequences e.g. waterlogging of over-irrigated lands, depletion of soil fertility, pesticide pollution, nitrate pollution of aquifers, and so on. Moreover, vast areas of India's countryside are left out of enhancement programmes. It is rightly said that 'Forty years of planned development has created an India in which islands of prosperity peep out of a sea of poverty' (*ibid.*: 34).

Since independence, about 3,000 big multipurpose dams have been built; consequently, about 3 million villagers, mostly poor tribal people, have been displaced, and the majority of them still live like refugees. After they move off their land, abandoning their livelihood and culture, they are forced to survive in new ways, which they had never imagined. The authorities closed their development files on these projects by providing meagre compensation for the losses that these people suffered. The benefits of large dams have accrued primarily to the industrialists and urban dwellers in general. The largest and the most controversial hydroelectric and irrigation complex, based on 30 major, 135 medium and 3,000 minor dams, and to be built over fifty years on the Narmada River in western India, was started in 1987 at an estimated cost of US$5 billion. The World Bank is the largest source of funds for this project. The dams will displace 200,000 people, submerge 2,000 km^2 of fertile land and 1,500 km^2 of prime teak and sal forest, and eliminate historic sites and rare wildlife. Thanks to the anti-dam movement by local and tribal people, referred to as India's first nationwide environmental protest, under the leadership of Ms Medha Patekar, the project has now become an issue of debate and dispute, and ultimately the work is not progressing optimally as proposed by the donor institution. In the light of uncertainty over government policies, rising pressure from the anti-dam movement and rejection of the major conditions proposed, the World Bank recently withdrew its support for this project. Internationally, the World Bank is having to rethink its policies, juxtaposing its role as a commercial 'development' institution with international demands for a greener and more socially aware lending body.

Through the process of capitalist globalisation, the Group of Seven (G-7) countries (the USA, Canada, Japan, the UK, Germany, France and Italy) promote a system of greater dependency on foreign loans and investments, the main sources of which are the World Bank and the IMF. Through the pressure of their terms and conditions, the receiving nation is bound to satisfy their objectives (*cf.* Farmer 1993: 145–146). Foreign companies and investments are now rapidly and easily capturing the Indian market while replacing the indigenous system, and behind the scenes they are also influencing the political system to gain favour. India's foreign trade deficit of US$1 billion in 1960–61 reached US$9.5 billion in 1990–91. This resulted in major changes to economic policies under the pressure of the World Bank. Moreover, to try to clear the foreign deficit, another loan was taken from the World Bank, ultimately creating a further economic threat. Consequently, India has to devalue its currency from time to time to suit the bank's conditions. It is charged that 'international debt has accelerated the rate of extraction and conversion of natural resources to enable the country to meet its external financial obligations' (Reed 1992: 143). The multinational companies of the G-7 countries and the World Bank are also promoting structural adjustment, mostly with the aim of securing their capital and gaining more profit on investment through the system of devaluing the Indian currency[1], promoting an open market and checking the nationalisation process. In the absence of (adequate) social objectives in adjustment programmes, structural adjustment has failed at different levels (*cf.* Reed 1992: 36–40).

In recent years, many non-governmental organisations (NGOs) and voluntary organisations have been established with specific objectives, primarily to try and activate local groups to play a watchdog role in preventing further ecological deterioration and to canvass public opinion in general for their active participation in development programmes. Self-help and self-reliance have been at the core of such programmes. NGOs are highly diverse organisations, are engaged in various activities and operate at a variety of scales. As watchdogs on the activities of governments, they play a key role in sustainable development, which perhaps governments themselves are unable to. The World Bank is also now encouraging alliances with NGOs which have traditionally been vociferous opponents of the World Bank, although of course the money granted is transacted through government channels. However, again the omnivores are slowly capturing the NGOs, through which they can extend their capital. Many NGOs are now being established in order to receive money from abroad. Initially, they follow their objectives, but over time the NGOs are mostly used for personal economic benefits. There are currently around ten thousand NGOs in the country. (This observation is based on various Indian newspaper reports during 1997–98. Through visits to and close interaction with the areas where NGOs are running their projects, one could gain such an impression.) Here the issue of moral ethics comes into being. If one lacks the sense of service as moral duty, no system can achieve its basic goals. There is a need for an attached system that may serve the role of watchdog in preventing NGOs from betraying their constitutions and objectives. This is also an example of cultural downfall where economic gain is the end.

The ideological search: Gandhi vs Nehru

Mahatma Gandhi (1869–1948) and Jawaharlal Nehru (1889–1964) were more responsible than any other persons for influencing the Indian psyche and the society for mass awareness towards an independence movement, and the basic framework of the philosophy of development.

[1.] The value of the Indian currency was Rs7.11 to the US dollar in 1970 but declined to around Rs42.00 in mid-1998.

Committed to non-violence (*ahimsa*) and self-realisation (*svachetana*), Gandhi wanted to solve India's problems from the perspective of individual conversion and with the ideology that 'every man has an equal right for the necessities of life even as birds and beasts there' (Segal 1965: 186). In pre-partition India, his vision was represented by 'religionisation of the political', directly related to the ultimate meaning and significance of human existence. However, in independent India his successor, Nehru, transformed it into a 'politicisation of the religious', referring to the tasks of political mobilisation and acquisition of power (Larson 1995: 199). Gandhi advocated doing the right thing at the right time through seer participation. According to him, the 'right thing' is a moral order (*dharma*) operated by right action (*karma*). Gandhi's emphasis upon self-realisation and rules of conduct and virtues is essential for spiritual life and also for the maintenance of the social order. For him 'right' comes only through the realisation of truth: he said 'God is the means and Truth is the end'. Truth and non-violence 'appear to him to be the same thing or different aspects of one and the same thing, and he uses these words almost interchangeably' (Nehru 1960: 279).

During the pre-independence period, Gandhi had attempted a public awareness movement through his walking tour, *padayatra*. The march was to spread the message of communal harmony and realisation of society's problems related to poverty and inequality. His disciple, Vinoba Bhave, continued this march to persuade landlords to donate land to the landless. In many areas it was successful; however, examples of donating rough and waste land were also recorded. This shows the lack of a conscience to support the poor.

Gandhi's successor, Nehru, was his apparent antithesis and wanted mechanised agriculture and rapid urban-industrial growth as the route to development. Segal (1965: 308) remarks that:

> Nehru never wished to be a tyrant, but he became one – not a great one but a petty one, and not through will but through vacillation. He reigned but did not rule, he commanded but did not conduct, he arbitrated where he should have resolved. ... He despised corruption and recognised its dangers, but he sat by, silent, while political colleagues, and even relatives, openly engaged in it.

The contrasts and contexts between Gandhi and Nehru can be summarised in tabular form (Table 3.1). Gandhi fought throughout his life to restore the old ethical values and for the revival of cultural identity. However, his successors and politicians interpreted him in their own way for their own benefit. In this way, the seed of self-interest, polluted political environment and sordid love for power was planted.

Until his death in 1964, Nehru had been prime minister of India; he 'was wrapped in visions of the future so tightly that he lost sight of the disorder and dismay everywhere around him' (*ibid*.: 309). With Gandhi's assassination in 1948, and the deaths of the leading voices of alternative models, Subhas Chandra Bose in 1945 and V.B. Patel in 1950, Nehru was free to develop his own ideas and rule the country according to his own consensus. Nehru's daughter, Indira Gandhi, using the populist rhetoric

Table 3.1 Gandhi and Nehru

Context	M.K. Gandhi	J.L. Nehru
social organisation	Indian village	urban society
industrial growth	handicraft, small level	heavy industries
change	individual, attitudinal	social and economic
democratic frame	moral democracy	social democracy
ideology	indigenous	Marxian
towards non-violence	alternative to class war	instrument of struggle

of 'removing poverty', became prime minister in 1971. Of course, she destroyed the boss structure of the old Congress Party, but she failed to replace it with an effective system. In 1984, Mrs Gandhi was assassinated, and following the 'dynastic succession' her airline pilot son, Rajiv Gandhi, was given the prime ministership. He was defeated in the general election in 1989, and was also assassinated, in 1991. By his death dynastic rule was ended. The lineage of father (Nehru) – daughter (Indira Gandhi) – and grandson (Rajiv Gandhi) held the prime ministership of India for thirty-nine years in the fifty-year period after independence. No leading party has ruled at the centre for a longer period since 1989, except from March 1991 to May 1996 (*cf.* Larson 1995: 228). After the downfall of Narasimha Rao's Congress government in May 1996, Mr Bajpayee (Bharatiya Janata Party, BJP) succeeded as prime minister, but for only thirteen days. He had been replaced for one year (1996–97) by Mr Devegouda, who was nominated by the United Front, an alliance of fourteen political parties having different ideologies, and with the support of the Congress Party, which pressed to replace the prime minister. That is how on 21 April 1997 I.K. Gujral took over the chair, replacing Mr Devegouda. Great political chaos was created by the Congress Party, which withdrew its support for the government. Therefore on 4 December 1997, the president of India dissolved parliament and imposed an unexpected burden of parliamentary elections (February–March 1998).

The government at the centre represents a combination of contrasting ideologies and varying ambitions of the politicians who are gathered under one umbrella to enjoy the lust of power. On 19 March 1998, a new BJP government was formed in alliance with nineteen other political parties; Mr Bajpayee returned as the sixteenth prime minister of independent India. Factionalism is the chief characteristic of Indian politics at central, state and regional levels. Tricks like electoral populism, emotional blackmailing of voters, manoeuvring and bargaining for support are commonly used by politicians to get a seat in the government – enjoying its fruits of power, influence and affluence. It has now reached its worst stage – every day some new scam or scandal is recorded. According to a newspaper survey, during the 1948–97 period the sum of major reported scams and scandals in India was around Rs20,428.7 billion (i.e. US$544.77 billion) – that is Rs21,503.8 (US$573.4) per person (*cf. Amar Ujala*, a daily newspaper from Allahabad, 4 February 1998: 7). In all these cases, politicians and ministers from the central and state governments were indicted.

In the 1998–99 Union Budget, the Indian government increased its allocation to the departments of Atomic Energy and Space Research by Rs13.66 billion in order to promote defence and nuclear power. This increase is more than five times higher than that for health, over 52 per cent higher than that for the central education vote, and over 72 per cent higher than the increased allocation for rural employment and poverty alleviation (*The Hindustan Times*, 9 June 1998: 22). So, the world's second most populous country, while propagating its ideology of 'truth and global peace', is in reality involved in a nuclear arms race, as evidenced by its nuclear test explosions on 11 May 1998 – ironically the Buddha's birthday. After all, the Buddha enunciated his philosophy of compassion, peace and brotherhood to the world in the sixth century BCE. This reminds one of the statement by the former prime minister, the late Indira Gandhi, in 1981 that the cost of building an intercontinental ballistic missile (ICBM) was equal to that of building 340,000 primary schools, or 65,000 health-care centres (*ibid.*). It was Indira Gandhi who ordered India's first atomic test, on 18 May 1974. These nuclear tests represent a kind of insanity that in no way promotes national prosperity but instead enlists the support of the poor masses through emotional blackmail for purposes of making India a nuclear power and thereby entering an ongoing struggle for hegemony through weapons build-up and terrorism. Yet it is the poor who ultimately bear the principal costs through their inability to meet their basic needs. Hence the rich become richer and the poor poorer – a development strategy devoid of sensitivity to moral ethics or brotherhood.

From rethinking to remaking

It seems certain that, as *Homo sapiens*, we should face the physical realities of the world, but also as human beings we should fulfil our psychological and spiritual needs. Only with a proper balance in the light of the situation will we guarantee a happy and creative future for the ecological world view. It is simple to be happy, but difficult to be simple! Above all, the country is in pursuit of a great social revolution – a revolution of mass consciousness to fight against corruption and the evils of society, to establish the symbols of honesty and working culture. In the recent past, we have lost the demarcation line of existence between right and wrong, moral and immoral, and also the realisation of the basic *dharma* (right action). This resulted in our acceptance of honesty as a psychic disease. The politicians and the bureaucrats are mainly responsible for creating this situation, but who will now remove it? A great social revolution with *svachetana* (Self-realisation) is suggested as the only strategy! However, again the basic problem is still unanswered. How is it to be implemented?

The core issues and conditions for an alternative, sustainable, development as identified by the World Commission on Environment and Development (1987) are population and development, food security, biodiversity and ecosystems, energy, industry, and the urban challenge. However, for direction and implementation, changes in thinking and motivation are a prerequisite. It is the people who *do* development, not the government, and therefore real development is ultimately a local activity and is generated by ethical issues. The neo-Marxist approach regards inequality as the major hindrance to development, and the latter is not possible within the existing world system. As in other developing countries, India's most crucial problem is a mass of absolute poverty, which is more dangerous than any other pollution! The present political and economic system is unable to solve the problems. Of course, no one can suggest a blueprint for eco-friendly development in the future; however, this does not mean that there is no prospect of a peaceful future.

Concerning the context of theoretical debate and perspective in India, postcolonialism is assumed to be a Eurocentric construct, post-traditionalism is too naive and postmodernism is yet more moral–idealist thought. At least some postmodern views of development convey a clear sense of moral duty or commitment (for a detailed critique and evaluation, see Simon, Chapter 2 this volume). Being part of the Indian grassroots, I support the sentiments of Lehmann (1997: 577): 'expert hegemony in which the poor of the world are turned into laboratory specimens, endlessly classified, catalogued and manipulated by a supremely arrogant and power-hungry professional community ... who survey the globe from an air-conditioned panopticon'. Above all, we have to cultivate the sense of self-realisation and the will to become directly involved to help the poor. Theory is the tool and practice is the result of inner quest; we need an integrated unity between them.

A system of moral education based on age-old ethics (religion in the sense of moral duty) and the use of indigenous methods needs to be promoted. Alas, this requires the strengthening of grassroots democracy based on reconstituting society into more homogeneous units. Proposals for a 'multi-religious state' (like biodiversity) are not a new idea. At the time of independence, C. Rajgopalachari (1959: 83, cited in Smith 1963: 151) proposed the idea in this way:

> And if India's government is to be an institution integrated with her people's lives, if it is to be a true democracy and not a superimposed western institution staged in Indian dress, religion must have an important and recognised place in it, with impartiality and equal reverence for all the creeds and denominations prevailing in India.

Larson (1995: 294) advocates: 'Perhaps this old idea from the time of independence, dismissed at the time as being naive and unworkable, should be given a renewed hearing as India approaches the half century mark of her independence. It is a sad point in intellectual history that 'after Independence, when the Indian ruling class modelled itself on its departed counterpart, any emphasis on the "flories of ancient India" came slowly to be viewed as an act of Hindu fundamentalism' (Chakrabarti 1997: 2). Further, in the light of metaphysical grounding in Indian thought, Deutsch (1989: 265) suggests that:

> we might yet find a way to bring our scientific understanding of nature's organic complexity into an integral harmony with a spiritual under-

standing of reality's simplicity. The *Gita* (IV, 18) states laconically: 'He who sees inaction in action and action in inaction, he is wise among men, he does all action harmoniously.'

In fact,

> Morally we ought, as best we can, to allow the living world, and the entities thereof, in their diversity, to thrive in richness, harmony and balance. In all things we must ask whether our actions are conducive to, or at least compatible with, the fullness and wellbeing of life.
>
> (Johnson 1991: 288)

Gadgil and Guha (1995: 124) propose an alternative development paradigm flowing out of 'conservative–liberal–socialism' and marching towards eco-friendly alternative development, which has the following elements:

1. Active public participation in democracy and bureaucracy at the various levels of power and transaction of resources.
2. The management of natural resources should substantially be devolved to the local population, taking into account the traditional (also indigenous) knowledge and sensitivity to sustainability in the frame of 'existence–maintenance–continuity'.
3. Decentralisation and mobilisation of resources and appropriate technology with the support of wider and more active public involvement.
4. Promotion of a more equitable society in its regional setting compatible with its natural and social/historical background or 'heritage ecology' (*cf.* Singh 1995).
5. Realise that the Indian experience, with its corrupt, inefficient and wasteful bureaucratic apparatus, is thoroughly discredited; the private and voluntary sectors (e.g. NGOs) are relatively much better and should be promoted in the service of society.
6. Economic enterprises should not be selected on the basis of appropriate scale, but rather on their validity, rationality and regional suitability.
7. The sense of tolerance and acceptability should be replaced by a deep sense of awareness and self-realisation (*svachetana*) in the context of reviving age-old ethics in the context of eco-justice, what Skolimowski (1990: 102) calls *reverential* development, where reverence, frugality and eco-justice form the minimal core of intrinsic values for right conservation and sustainable development.
8. Government policy should be integrated with the people's lives and ideologies, i.e. the morality, ethics and faith (*dharma*), and should not be a superimposed, Western institution staged in Indian dress, but have religious faith with impartiality and equal reverence for all that the early leaders like Rajagopalachari and Radhakrishnan had provoked during the 1950s.

It seems that by mobilising politicians and bureaucrats to adopt a more holistic view of the global ecosystem and by demanding the maintenance of deep human issues, many of the intricacies and obstacles can be addressed. The need for spiritually oriented education should also be accepted as an alternative strategy drawing upon India's rich variety of indigenous spiritual and cultural heritages. Like the Gaian concept of 'the alive Earth', the realisation of the sense of deep connectedness is the concern of deep ecology. One of the most sacred ancient texts, dated *c*. 2000 BCE, the *Rig Veda* (X.18.11) mentions a funerary hymn honouring the mother Earth (*cf.* Singh 1995: 192):

Heave thyself, Earth
nor press thee downward heavily:
afford him easy access
gently tending him.
Earth, as mother wraps her skirt
about her child, so cover him.

Following the ecological notion that everything is interconnected, ecological awareness leads to the ethical question of finding an inner guide for right action, right thought and sustainability. Devereux (1996: 16) says that: 'It is our fate, not the fate of the planet, that is in question. We have to save ourselves. We need stewardship for our species. Rather than thinking of healing the Earth, we should be looking to the Earth to heal us.'

Remember also that modern education has alienated humankind from the cosmic integrity because most of it is imparted in closed buildings and laboratories (Asopa 1994: 87). Carl Jung has acknowledged spirituality as an integral aspect of human nature and a vital force in human life (Capra 1983: 406). Capra (*ibid.:* xix) feels that spirituality in its ultimate nature will lead to profound changes in our social and political structures. He further adds that: 'we find a profound state of world-wide crisis. ... It is a crisis of intellectual, moral, and spiritual dimensions; a crisis of a scale and urgency unprecedented in recorded human history' (*ibid.:* 1).

Schroeder (1996: 305) believes that by broadening our world view it is possible for us to take the spiritual values of nature seriously without abandoning our scientific heritage. Let us promote the attitude to human-kind (say moral ethic) by balancing and reconciling the rational–scientific and the intuitive–spiritual viewpoints in ourselves and our society. Ethics is a technique of the soul and a seeking to understand a deeper nature of things, especially why we should behave in one particular way rather than another (Skolimowski 1990: 97).

Does economic growth lead to development in India or other countries with a long tradition of particular value systems? Can this be called ho-listic or sustainable development? Certainly not! Development denotes progress and improvement in the quality of life and human peace. In a deep sense, this kind of development can be achieved only by a balanced integration at various levels in all spheres, including economic, social, cultural, scientific and technological, and moral and spiritual values. We need an integrated framework of sustainable development that combines contemporary ethical imperatives with traditional ethical codes; it should attempt to serve all people of all cultures, and promote the establishment of a place between humankind and nature (*ibid.:* 103). The fact that it may be difficult to implement in practice in no way negates its importance and desirability (*ibid.:* 102). Mass awakening, what Gandhi exemplified by his *satyagraha* movement, is 'the perfect example of how one could confront an unjust and uncaring, though extremely superior power' (Dwivedi 1990: 211). In contemporary India, the Bishnois (a small community in Rajasthan, western India, who practise a religion of environmental conservation), Chipko (the movement to hug trees to keep them from being cut down for industrial uses, started in 1974 by Sundar Lal Bahuguna) and Appiko (the Chipko movement in Karnataka, south India, begun in 1983) people are engaged in a kind of 'forest *satyagraha*' today. Of course, it is a hard task for those from that society, but not impossible. The Hindu view of cyclic incarnation has immediate consequences for conservation, moral ethics and respect for harmonious life, because all lifeforms have essentially the same soul (*atman*) which passes in cyclic form from one life form to another, thus completing 8,400,000 species (Palmer 1990: 53).

Following the analogy of the Buddhist 'middle path', India has to strike a balance between its deeply rooted value system to generate a cultural ethic in the service of national identity and to cope with the influx of postmodernist and postcolonial pressure, especially the techno-economic– ethical threat coming from the West. A synthesis between conservative 'moral imperative', liberal 'democratic capitalism', and 'socialist pattern of equity and empowerment' may be taken as an alternative development strategy. Moreover, regardless of the question of which path of development should in fact be followed, the government has to realise that in future the open-market economy and liberalisation would also have dangerous effects on the country's progress in respect of its resources, management and culture. A journalist has written provocatively:

> The neo-liberal policies, called 'structural adjustment', would not benefit the poor in the short or long run. The whole logic of adjustment is to transfer resources from non-tradable to tradable, unorganised to organised, poor to rich. This is how the 'free' market works – free to morality, equity, justice or concern for the poor.
>
> (Bidwai 1997: 12)

Let the Indian tradition of tolerating, assimilating and synthesising diverse strands flourish in making India a nation of brotherhood, peace and prosperity with equity.

> Since the roots of *ecological* trouble are so largely religious, the remedy must also be essentially religious, whether we call it that or not. We must rethink and refeel our nature and destiny.
>
> (White 1967: 1207)

Ethical value is the moral force in the sustainable existence, progress, maintenance and continuance of human beings. Gandhi's life and work do hold possible implications for postmodern approaches to the environment, mostly based on his principles of *ahimsa* and *aparigraha* (non-possession). He said:

> I must confess that I do not draw a sharp or any distinction between economics and ethics. Economics that hurt the moral well-being of an individual or a nation are immoral and, therefore, sinful.
>
> (Gandhi 1959: 33–34)

His theory of *ahimsa* was not strict like a sectarian rule; he said: 'whoever believes in *ahimsa* will engage himself in occupations that involve the least possible violence' (*ibid.*: 35).

Without any political propaganda, Acharya Tulsi's Anuvrat Movement has been promoting the process of self-purification and self-control, in many ways akin to Gandhianism, since 1949. In support of this movement, Radhakrishnan (quoted in Chapple, 1993: 62) said:

> A civilized human being must be free from greed, vanity, passion, anger. Civilizations decline if there is a coarsening of moral fibre, if there is a callousness of heart. ... It requires strict adherence to the principles of good life.

For Raine (1997: 7, 11) India is not a country; it is a state of mind which gives birth to beauty and embodies the spirit:

> the treasury of spiritual knowledge and practice is beyond doubt India's greatest resource. And it may be that our bankrupt materialism has brought us in the West to a point where we recognize our own need to relearn what our full humanity entails.
>
> (*ibid.*: 11).

Think universally, see globally, but act locally. This is an appeal for cosmic vision, global humanism, and self-realisation (see Singh 1999 for details). Should we call it 'development'?

Acknowledgments

I personally express my gratitude to David Simon, who put forward so many constructive criticisms and also kindly thoroughly edited and polished the English. Our friend, Anders Närman, too has been supportive in the development of this chapter in several ways at different stages.

References

Asopa, S.K. (1994) Environmental ethics: a Hindu perspective. In Mishra *et al.* (eds) *op. cit.*: 69–89.

Bidwai, P. (1997) Rich village, Poor village. Neo-liberal policies have hurt rural India and its poor. *The Times of India* (Sunday Times), Delhi, 12 January: 12.

Bose, A.N. (1997) India in the 50th year of Independence: economic situation. What happened? What would be? *Samayika Varta* (a Hindi quarterly) **21** (July–September): 20–59.

Capra, F. (1983) *The Turning Point*. Fontana, London.

Chakrabarti, D.K. (1997) *Colonial Indology. Sociopolitics of the Ancient Indian Past*. Munshiram Manoharlal Publishers, Delhi.

Chapple, C.K. (1993) *Non-Violence to Animals, Earth, and Self in Asian Traditions.* SUNY Press, Albany, NY.

Deutsch, E. (1989) A metaphysical grounding for natural reverence: East –West. In Callicott, J.B. and Ames, R.T. (eds) *Nature in Asian Traditions of Thought.* SUNY Press, Albany, NY: 259–265.

Devereux, P. (1996) *Re-Visioning the Earth. A Fireside Book.* Simon & Schuster, New York.

Dwivedi, O.P. (1990) Satyagraha for conservation: awakening the spirit of Hinduism. In Engel and Engel (eds) *op. cit.*: 201–212.

Elliott, J.A. (1994) *An Introduction to Sustainable Development.* Routledge, London.

Engel, J.R. and **Engel, J.G.** (eds) (1990) *Ethics of Environment and Development.* University of Arizona Press, Tucson.

Farmer, B.H. (1993) *An Introduction to South Asia* (2nd edn). Routledge, London.

Gadgil, M. and **Guha, R.** (1995) *Ecology and Equity: The Use and Abuse of Nature in Contemporary India.* Routledge, London.

Gandhi, M.K. (1959) *My Socialism.* Navajivan Publishing House, Ahmedabad.

Johnson, L.E. (1991) *A Morally Deep World.* Cambridge University Press, Cambridge.

Kothari, R. (1990) Environment, technology and ethics. In Engel and Engel (eds) *op. cit.*: 27–35.

Larson, J. (1995) *India's Agony Over Religion.* SUNY Press, Albany, NY.

Lehmann, D. (1997) An opportunity lost: Escobar's deconstruction of development. *Journal of Development Studies*, **33**(4): 568–578.

Mishra, R.P. *et al.* (eds) (1994) *Environment, Development and Education.* Heritage Publishers, New Delhi.

Naess, A. (1989) *Ecology, Community and Lifestyle. Outline of Escosophy.* Cambridge University Press, Cambridge.

Nehru, J.L. (1960) *The Discovery of India.* Oxford University Press, Delhi. Reprinted.

Orwell, G. (1945) *Animal Farm.* Secker & Warburg, London.

Paden, W.L. (1992) *Interpreting the Sacred.* Beacon Press, Boston, Mass.

Palmer, M. (1990) The encounter of religion and conservation. In Engel and Engel (eds) *op. cit.*: 50–62.

Radhakrishnan, S. (1956) Introduction. In S. Abid Hussain, *The National Culture of India.* Jaico, Bombay: vii–viii.

Raine, K. (1997) All roads lead to India, *Resurgence* **183** (July/August): 7–11.

Rajgopalachari, C. (1959) The place of religion in future India. In *Message of India*. Government of India, New Delhi, vol. I: 83; cited in Smith (1963: 151).

Reed, D. (ed.) (1992) *Structural Adjustment and the Environment*. Earthscan, London.

Schroeder, H.W. (1996) Spirit of the forest. In Swan and Swan (eds) *op. cit.*: 294–306.

Segal, R. (1965) *The Crisis of India*. Penguin, Harmondsworth. Reprinted in India: Jaico, Bombay, 1968.

Simon, D. (1999) Development revisited. Chapter 2 in this volume.

Singh, K. (1990) Brahmin power, *Sunday* (23–29 December): 19.

Singh, Rana P.B. (1995) Heritage ecology and caring for the Earth, *National Geographical Journal of India* **41**(2): 191–218.

Singh, Rana P.B. (1996) The Ganga river and the spirit of sustainability in Hinduism. In Swan and Swan (eds) *op. cit.*: 86–197.

Singh, Rana P.B. (1999) Ethical values and spirit of sustainability in Indian thought. In, F.J. Coasta *et al.* (eds) *Geographic Inquiry at the Beginning of New Millennium*. Westview Press, Boulder, Colo.

Singh, Rana P.B. and **Singh, R.B.** (1981) *Changing Frontiers of Indian Village Ecology*. National Geographical Society of India, Varanasi, Research Publication No. 27.

Skolimowski, H. (1990) Reverence for life. In Engel and Engel (eds) *op. cit.*: 97–107.

Smith, D.E. (1963) *India as a Secular State*. Princeton University Press, Princeton, NJ.

Swan, J. and **Swan, R.** (eds) (1996) *Dialogues with the Living Earth*. Quest Books, Wheaton, Il.

Watts, A.W. (1962) *The Joyous Cosmology*. Random House, New York.

White, Jr, L. (1967) The historical roots of our ecological crises. *Science* **155** (10 March): 1203–1207.

World Commission on Environment and Development (1987) *Our Common Future*. Oxford University Press, New York.

CHAPTER 4

Between theory and reality
The area specialist and the study of development

Michael J.G. Parnwell

Preamble: nine actors and me[1]

Thailand

Khun Prasert – villager and owner-operator of a diamond-cutting workshop, northeast Thailand: Khun Prasert was until recently riding the roller-coaster of the 'diamond' (zirconium) and gemstone industry in northeast Thailand. He jumped on the thundering bandwagon in 1990, a few years after the industry's boom started. From absolutely nothing in 1985, the industry had risen like a phoenix from the ashes of a depressed regional economy, benefiting from and contributing to the dynamism of one of Bangkok's strong export-oriented industries, cosmetic jewellery, and creating a rare and powerful alternative to rural out-migration. For a time it represented a classic example of 'bottom-up development' (see Parnwell and Khamanarong 1990, 1996). Everybody wanted a slice of the action, either as workshop labourers or, if a bit of capital could be begged or borrowed, as budding 'diamond' entrepreneurs. The state was nowhere to be seen, and many people remarked that the industry was particularly successful *because* of this. Ten years later the industry collapsed: too many cooks had spoiled the broth, and taken the glisten out of this sparkling industry. Many people, like Khun Prasert, ended up no better off than when the industry had started – a binge of immediate gratification left no capital and thus no alternative to the industry other than a return to either farming or Bangkok. Workers were left to rue the costs of abysmally poor working conditions – seriously damaged eyesight, stumps where nimble fingers had once played – which were traded off against the useful short-term gains that this remarkable rural industry had once offered. As Khun Prasert said, ruefully, 'if only someone – the state, an NGO, a businessperson, someone – had regulated this industry's development so that

---

[1]   These brief case studies are all based on real-life encounters, although the precise identities of the individuals concerned have been masked. The case studies are all relevant in some way to the objectives of this chapter. The fact that all the people featured in these cases are men is probably quite significant.

we might have had a sustainable alternative to farming and migration in the region'.

Khun Somboon – headman, Ban Sa'aat village, northeast Thailand: I sat there one long afternoon in the early 1980s, writing up my field notes in the shade of Headman Somboon's grand compound. He sat opposite me on the bamboo bench, feverishly engaged in some writing of his own. He was filling in some forms, heaps of them: adding words, figures, and signatures – eight of them, all different. The forms related to work undertaken in Ban Sa'aat under the government's then Rural Job Creation Scheme – a nationwide programme designed to create work in rural areas during the slack dry season, while simultaneously helping to build or maintain vital village infrastructure. Ban Sa'aat village was constructing a huge, communal fish pond, and Headman Somboon had, apparently and unusually, put hundreds of hours work into this valuable project. In reality, Headman Somboon had simply put a few minutes' work into 'creative writing' but had managed to cream off at least 25 per cent of the village's entire development budget for the summer. Here it was, corruption in action, and I sat passively by. What right did I have to pass judgement; get involved?

Ajaan Amnuay, lecturer, Northeast Thailand: Ajaan Amnuay specialises in rural development in a social sciences department within one of the region's universities. Although he comes from elsewhere in Thailand, he speaks the regional dialect, understands the local people, and is not averse to getting his feet, or at least his tyres, dirty. As a lecturer, he has a reasonable degree of status not only in the villages but also when dealing with various government officials, which his research, and his job, requires him to do. He has some fresh ideas about how development in the northeast might be promoted. He has just finished studying at an English university, and he has been quite taken with notions of alternative, appropriate, participatory and grassroots development. The officials listen attentively, but what's in it for them? Mustn't be too radical, mustn't rock the boat – got a career to think of – unless I could perhaps sell these ideas to my boss, who might then champion himself as the 'friend of the people'. No, too risky that – not sure that head office would approve. Best to revert to rhetoric at this point – say all the right things, do nothing. After a while, Ajaan Amnuay gets a little tired of banging his head against a teak wall, and anyway, he has got that new house to pay for and his own career to think about. The headship of department is up for grabs: a bit more power, clean tyres and a go-fast stripe – that will do nicely.

Sarawak, Malaysia

Henry Sagoo, state planner: Henry is from one of the myriad ethnic groups indigenous to Sarawak, and as such should find it quite easy to identify with their current development plight. The Bidayuh, like so many of the state's societies, are facing very rapid change to their environment, their economy and their socio-cultural identity. Rapid modernisation, intensive deforestation, deep capitalist penetration and growing out-migration are all taking their toll on the rural people of Sarawak. Henry attends seminars and conferences all around the region, and sometimes around the world, and is thus quite well imbued with prevailing development debates and paradigms. In any case, he received his doctorate in development economics and planning in the United States. But while the current trend is towards small-scale, flexible, participatory and compassionate forms of development, Henry toes the national party line, enunciated in Kuala Lumpur, which still thinks big (and tall, and mega) is best, most efficient and effective, and that top-down should be top-dog. He talks with representatives of the people, but seldom listens. He attempts to convince them that the prevailing orthodoxy – essentially, turn the whole of Sarawak into a plantation: it worked for West Malaysia – is the most pragmatic path. Anyway, if you do not agree you will not enjoy the munificence of the state. If you wish to protest, you will be seen as attempting to destabilise the state, which is rather risky. Henry is a practitioner, but he understands politics and patronage very well. If you want to get on, then swim with the current and not against it. Furthermore, your people will be proud that one of them has risen to such heights of bureaucratic office.

Kamarau, Iban lad: I watched Kamarau hacking away at a bushy rattan plant (*wii duduk*) with his *parang*. Perhaps seven years' growth was terminated in an instant. The reward, a small stem core the size of a carrot, which, with three others (four big bushes, fourteen minutes), would add a little texture and flavour to that evening's meal. 'The rainforest is our supermarket,' he said – crash and carry. Kamarau is a typical Iban lad – immensely knowledgeable about the rainforest ecosystem, but most happy when chasing women in town. He is in and out of work these days, but not so long ago he had worked the logging camps and had so much money that it would not fit in his pockets. The felling of timber was good for him, and also therefore the state. His future was in town anyway, so what use was the forest? He would send his mum some money when he had it, but has not been able to do this very often lately since the logging camps started closing down. He had not saved anything during the good times, and now there is so much competition for work that his prospects of owning his own house in town look very bleak. Even the women are less interested in him, the epitome of Iban manliness. Sure, there is some factory work about, but they are offering such paltry wages that it is better for the Indonesians from across the border to do it. In Kamarau's words: 'sustainable development – that's about the future, isn't it? I may be dead tomorrow – what about now?'

Abdul Kamaluddin, lecturer: Abdul finished a PhD in development sociology at a university in England and then joined the new state university in Sarawak. He smokes a pipe and quotes from Shakespeare. He is well versed in development theory, but most of his thesis was concerned with development reality. He studied communities who had been transformed from shifting cultivators to plantation workers almost overnight. As such, he was grateful to state policy for giving him a development problem to get his teeth into. Abdul's thesis had a couple of theoretical chapters followed by a good array of empirical material. One of the examiners remarked that the theoretical and empirical material were never very closely integrated. Abdul's defence was that he finds difficulty in reconciling the two. Dependency theory did not seem to fit the Sarawak case particularly well, and yet there was nothing much else that he could use that was tailor-made for his study. There were no indigenous theories of societal change that he could use; and, as an area specialist, he felt that his most valuable contribution to the furtherance of knowledge would come from the insight that he – as a local indigene – was able to provide through his case studies, rather than by trying to advance a fresh theoretical formulation or two.

Since taking up his lecturing appointment he has enjoyed a meteoric rise up the academic ladder, as he was the only staff member at the time who had been educated at a Western university. In this sense, his background and training are very useful to him although, unfortunately, since becoming dean he has not been able to keep up his research, apart from doing a few 'consultancies' for the government. He is, however, very keen to facilitate collaborative research with his Western colleagues.

Indonesia

Pak Kotawali, municipal office, Sulawesi: How to obtain research support? I am not sure quite how it happened, as on occasions I find Indonesia's politico-cultural institutions impenetrable, but I succeeded in obtaining a research support grant of Rp40 million (alas only £10,000, although it sounds more) from the mayor's office for a project on small-enterprise development in urban areas. For the mayor, this was, presumably, politically expedient – the support of small-scale industries has recently been given a high priority by the Indonesian government, and so too has the promotion of the hitherto backward eastern isles of the archipelago. Engaging a Western scholar in such timely and topical research would be bound to send the right signals to Jakarta. Events thereafter revealed how hollow this support actually was.

Two years later, we are still trying to prise out of his office the 75 per cent of the grant that is still outstanding, and which has already been 'spent' on enumerators' fees. Nobody has the slightest interest in closely studying our research findings, let alone acting on our recommendations. To a large extent, the findings and recommendations run contrary to the municipal administration's own vision of the future: where we see a role for the micro, the officials appear hell-bent on promoting macro solutions to development problems; where we seek to understand, respect and incorporate cultural and social parameters into our policy prescriptions, the officials are wholly preoccupied with economism; where we have sought to emphasise the holistic interconnectedness of activities, problems and solutions, officials persist with a narrow focus on individual sectors and aspects. Clearly, they expected, or wanted, us to tell them what they hoped to hear – that their present actions were just what was needed – and not that they might consider reorienting their policies and programmes. So, the classic reaction is to do nothing. The political capital has presumably already been acquired; and anyway, an election is looming, so it is best to hold fire for a while.

Pak Pramono, academic practitioner, Sulawesi: Pak Pramono was my comrade-in-arms on the small-enterprise development project. He is a very bright and committed individual, and his main responsibilities focus on the provision – via the Department of Small-Enterprise Development – of training, advice and support to small-scale and petty entrepreneurs from the city's industrial and service sectors. Pramono had started a doctoral programme in the 1980s but never quite finished, his onerous duties as a lecturer, trainer and administrator stifling any real opportunity to fulfil his scholarly ambitions. Pramono does most of the donkey work, while his immediate superiors lead a stress- and largely work-free existence. They take a great deal of any credit that emerges from Pak Pramono's work, and this even extends to dividing up any research income that filters through to the department – approximately 40 per cent of the research grant referred to above was (or at least theoretically will be) syphoned off by up to ten of Pak Pramono's senior officers, as is 'normal' practice. They also become (passive) participants in the research project, one consequence of which is that the final report must be approved by all of them – irrespective of their lack of involvement in or even comprehension of its findings – which in turn means that anything even remotely critical, controversial or original is almost certain to be toned down. Here, the academic system functions as an integral element of the political community within which it is placed, and it is dangerous to stick one's neck out. Accordingly, Pak Pramono has learned to operate within the system – attempting to make a difference to the city's disadvantaged population without rocking the boat unduly, and remaining philosophical about the *status quo*. In such a way, the agenda set at the top (and this is typically self-serving) tends to permeate all areas and sections, while ideas, initiative and idealism at the policy/action interface tend to dissipate with depressing frequency.

Haji Ibrahim, owner of small-scale rattan workshop, Sulawesi: Haji Ibrahim is one of the small-scale entrepreneurs upon whom our study focused. In recent years he, like scores of other rattan furniture producers, has felt the pinch caused by the opening of a large-scale capital-intensive rattan factory on the edge of the city. This factory was promoted and supported by some of the city's leading figures, several of whom have a stake in its operation. The arrival and relative success of this factory has been achieved at a cost to the city's small-scale operators, who have seen a dramatic increase in competition for (and price of) raw materials, labour, capital and markets. 'Unfair!' they say, albeit fairly quietly. Tough! In order to seek to protect their interests, the rattan producers have formed themselves into a trade association, which gives them some bargaining power in acquiring basic materials and capital, and helps with the regulation of prices and inter-firm competition, and with the securing of markets and customers. Haji Ibrahim is quite high up in the executive of this association, due at least in part to his high social standing on account of his pilgrimage to Mecca. He has also been able to use his position at the head of the association to enhance the performance and prospects of his own firm, which has developed apace in recent years in spite of the sector's overall poor performance. 'Unfair,' mutter some of the other members. Tough! Meanwhile, a number of petty rattan producers have not been allowed into the association, either because they lack formal operating licences or because they lack contacts within the industry. They are the ones who have felt the squeeze most of all, clinging on precariously for survival. They seldom cry 'foul' – it is an historical mess and they have no right to expect better, and in any case it will do them little good. Their long-term aspiration is to ascend the industry's ladder – at least then there will be someone below to exclaim 'tough' to.

Hull

And me: I like to get my feet dirty and want to make a worthwhile and meaningful contribution to development in my region of specialisation, Southeast Asia. My motivation in this regard started with my fieldwork in the northeast of Thailand, during which I lived in several rural communities. I think that this and subsequent such experiences, coupled with a certain familiarity with two local languages and a willingness and ability to immerse myself (up to a point) in local cultures and to try to see development realities through local eyes, has helped to shape the way that I perceive 'development'. Added to this, I have been associated for the last fourteen years with an area studies department which has given me considerable exposure to academic disciplines other than geography (where I started out), and the luxury of dedicating my career to the study of one particular geographical region.

I am not quite sure why, but intellectually I have always been much more comfortable with development realities than development theories. Somehow the latter always seemed to lack the power and ability to explain what I was constantly seeing and experiencing on the ground. I do not dispute the need for wider views and structures to give some shape and purpose to the study of development, and indeed have attempted to engage several theoretical discourses and frameworks in the course of research work that has taken in Thailand, Malaysia, Indonesia and China. However, I believe very strongly that development is about improvement, and I struggle to find concrete examples where detached, abstract, grand theorisation has helped significantly to improve things where it really counts – in the backyards of real people's lives. Recent advances in poststructuralism, postcolonialism, postmodernism and neo-populism have allowed me to come out from being a 'closet ground-truther' and, as we shall see in this chapter, to look much more quizzically at claims from my disciplinary counterparts that area studies represents an inferior branch of academe and a career *cul-de-sac*.

In spite of my predilection for eclecticism and for working close to the grassroots level, I do not think I have ever quite managed to achieve my objective of making an effective contribution to the process of development, as opposed to the study of development. I have never (presumably *mea culpa*) managed to channel my research findings and ideas adequately to those who have the power or responsibility to help to improve conditions for the social groups with whom I have been concerned. I find that, as my career has progressed, I have in any case become gradually further removed from 'independent' grassroots research but have instead increasingly worked with and through institutions in my host countries – and these institutions (academic, governmental, etc.) have seldom been those that, effectively, can or do make much of a difference in a development sense. I write journal articles and books that seldom venture outside academic circles. And so I find that my principal justification for the work that I do in the field of development increasingly relates to the indirect impact that education – opening people's minds – might have on friends and colleagues from the part of the world with which I am concerned. Given the crescendo of calls for the 'indigenisation' of development research, I find it ironic that many of the Southeast Asian students that I teach and supervise receive their first exposure to 'new development' thinking when they are in the United Kingdom. Western academic imperialism has had such a powerful influence on their own societies that they often have to venture outside to learn how to think for, and be, themselves.

Objective

This chapter claims a potentially pivotal role for the 'area specialist' in the new development studies. Because of the area specialist's ability to get close to the contextual realities of development, and to comprehend development issues in an unbounded, holistic manner, it is argued that this person is in many senses properly equipped to rise to the challenges that have arisen from the recent rethinking of development, and most particularly the emergence of poststructuralist, postcolonialist, postmodernist and neopopulist discourses. Such a claim strikes at the heart of an orthodoxy that privileges bounded, disciplinary approaches to the study of development, reifies theorisation while simultaneously denigrating 'empiricism', and where the dirty-footed 'fieldworker' has conventionally been considered the poor cousin of the ivory tower sophisticate.

This orthodox viewpoint has recently come under serious and sustained attack from several quarters as questions have increasingly been asked about the utility, relevance and validity of metatheory, the Western domination of development knowledge, and the patent failure 'of the hundreds of thousands of words written about development' (Edwards 1989: 135) to make an iota of difference to the plight of the poor and destitute. Alternative approaches are increasingly being advocated which privilege the micro over the macro, the particular over the universal, the indigenous over a hegemonic Western, contextuality over exogamy, actor over structure, and practice and/or reality over theory. While many quite naturally contest this pendulum swing between epistemological extremes, it has arguably created both more space and a greater need for development scholars who are capable of reading, understanding, explaining and contributing to the reality of development. It is the central contention of this chapter that the 'area specialist', while rarely possessing *all* the qualities that are required of the 'new developmentalist', has a potentially significant role to play in the future of development studies. However, orthodoxy may not be expected to cede ground readily, and thus area specialists who herald a grounded, informed, contextualised and interdisciplinary understanding of development realities have still to overcome many of the prejudices that are associated with the established structuring of knowledge, particularly that occurring within academic disciplines and revolving around a top-down approach to the formulation of theory.

The following discussion will first outline some of the changes that have recently been taking place in the field of development studies, before attempting to make a case for areal specialisation within the framework of the 'new development'. Questions of definition will be dealt with after the case has been established.

Problems and prognoses

In an important edited volume, *Rethinking Social Development: Theory, Research and Practice* (1994), David Booth and a number of leading development scholars attempted to take stock of the long march of development studies from the vantage point of the halfway house of a post-Marxian impasse (see also Schuurman 1993). Where had the path led, and where was it leading? What had the march achieved, and what might it achieve? What should it achieve, and how might it achieve this? Few answers presented themselves, but at least the asking of these quite fundamental questions invited, or more accurately reflected, a bout of navel contemplation within the development 'community'.

Since the mid-1980s, the world of development studies has moved into a new era. Past theorisations of development have increasingly been criticised as inappropriate and misleading, for reasons of massive overgeneralisation and abstraction, ideological blinkering, and the sometimes wilful dismissal of heterogeneity in the real world of development (Booth 1994: 3). The predominance of economism and unidirectional determinism had meant that the essentially human dimensions of development had either been reduced to representing components of systems, or overlooked altogether (*ibid.*: 5). The bankruptcy of neo-Marxist formulations (a contention that is strongly challenged by Corbridge (1993)) coincided with a growing distrust of overarching 'grand theories' of development, an increasing awareness of and preoccupation with diversity, and a parallel fragmentation of approaches to the study of development (Booth: 1994: 3). Difference, particularism, independent variability, polyvocality and multiple choice of development paths have moved steadily to the forefront of the field. Much greater attention is now paid by what Corbridge (1993: 450) calls 'development activists' (my 'new developmentalists') to the actual workings of development rather than an abstract representation of development processes (Booth 1994: 11). In some ways, these changes have brought the conceptualisation of development closer to the reality of development, but their pace, nature, extent and faddishness are rued by Booth because of their almost anarchic lack of coherence and collective direction. It is part of an individualising and anti-theoretical bent (*ibid.*: xiii), where empiricism is glorified, and where 'method is all and theory is nothing' (*ibid.*: 14). The end product is, or threatens to become, a 'mish-mash' of ideas and approaches, the overarching contribution of which is less than the sum of its constituent parts (*ibid.*: 12).

Both Booth and Corbridge concede the need to refine the theorisation of development in order to take account of criticisms that have emanated from postmodernist, postcolonialist, poststructuralist and neopopulist quarters. However, the theorisation of development remains essential, to Corbridge (1993: 467) because the growing interconnectedness of a globalising or interdependent world system requires a wider vision of social and economic transformation, and to Booth (1994: 14), more contentiously, because 'the generation of theories ... remains the ultimate objective of social science research'. The challenge is to find ways of generalising and thus theorising diversity in order to link 'development theory and practice, the local and the extralocal, the specific and the "universal"' (Corbridge 1993: 451) and to make some conceptual sense of what is happening (Booth 1994: 14): 'to change the world it is first necessary to offer an account of the world' (Corbridge 1993: 469).

Others, while rarely challenging the need for theoretical structures *per se*, have continued to press for alternative ways of conceiving development. Some claim the need for theory to be built from the bottom upwards, based fundamentally upon the contextual realities of development. Others challenge the relevance and value of many orthodox forms of development theory *and* practice, and argue the need for the development community to strive more consciously to make a positive contribution to the amelioration of pressing and enduring development problems. Yet others question the predominance of Eurocentric visions of development and the Western domination of the development industry, calling instead for a greater indigenisation of both. Because of constraints of space, I shall look briefly at the overlapping work of three authors (Brohman 1996; Edwards 1989, 1994; McGee 1995) who have sought to champion each of the above approaches. In the process, I aim to show how the perspectives they represent offer scope for area specialists to make a more central contribution to the field of development than has tended to be the case hitherto.

John Brohman makes the case for a form of 'popular development' that, in many respects, represents the antithesis of an increasingly maligned orthodox development. In the process, he makes a number of statements about the role of the development *researcher* that have a significant bearing upon the central objective of the present chapter. Brohman (1996: 324) favours 'a broader, more flexible vision of development capable of addressing diverse Third World realities' and claims that 'solutions to development problems must be sought in the contextuality of development, which is a product of particular historical processes ... and the subjective concerns of different groups of people' (*ibid.*: 325, 329). He argues that a '[g]reater familiarity with local experiences might produce more useful and applicable concepts, more appropriate methods, and more realistic expectations of the people involved in the actual work of development' (*ibid.*: 326). Theory building should therefore be based upon the development experience itself, and should thus incorporate a bottom-up directionality of theory formulation (*ibid.*: 352). The problem is that, within the orthodox framework, theorists and experts tend to be too far removed from the 'real world' of development. As a counter to this, Brohman argues the need for development workers to become more open to difference, and to be willing and able to learn from other societies and cultures, to take an interest in local knowledge and cultural practices, and to use these as the basis for redefining their development approaches (*ibid.*: 337–338).

If we accept such an approach as a realistic way forward, both in terms of representing a way out of the current theoretical and methodological impasse, and in respect of bringing development work closer to an effective engagement of the needs of 'distant strangers' (Corbridge 1993),[2] then clearly we must be looking for a rather different kind of 'developmentalist' than has dominated the scene hitherto. If contextual, moral and ideological detachment are part of the problem, and contextual familiarity, engagement and understanding an important part of the solution, then it follows that to be effective the 'new developmentalist' needs at the very least to be able to handle and privilege contextuality to a much more significant extent than has hitherto been achieved. This is where I believe the area specialist has an important, indeed central, role to play. In its purest form, areal specialisation requires deep contextual immersion, an ability to communicate with and intuitively to understand other societies and cultures – if not to think like 'distant strangers', then at least to comprehend how they think and act, and why. Only from such a localised and contextualised platform can an adequate foundation be established for the aggregation of development experiences that leads to the bottom-up conceptualisation of development and theorisation of diversity (Booth 1994: 27).

The second argument concerns an apparently widening gulf between academic research and the worlds of policy and action, a lack of 'relevance' in development thinking, and a research agenda that is driven by the needs of the researcher rather than those of the researched. Northern researchers are accused (Edwards 1989: 124) of using the Third World as a laboratory for social, economic and political experiments, turning development into a 'spectator sport, with a vast array of experts and others looking into the "fishbowl" of the Third World from the safety and comfort of their armchairs'. Similar sentiments are echoed by Lehmann (1997: 577) when he writes of 'expert hegemony in which the poor of the world are turned into laboratory specimens, endlessly classified, catalogued

---

2. It is important to point out that much of the discussion contained within this chapter relates to 'outsiders looking in' on the development experience of others. The logical continuation of many of the arguments that are presented here, most particularly the call for the 'indigenisation' of development research, is that the 'new developmentalists' should ideally be of and from the societies in question. Be this as it may, it does not eliminate one of the most important realities of development – that of the almost hegemonic domination of the field by Northern 'experts' and 'specialists'. As such, much can be gained from a critical appraisal of the attitudes, agendas and approaches of the development outsider.

and manipulated by a supremely arrogant and power-hungry professional community ... who survey the globe from an air-conditioned panopticon'. The development community sticks rigidly to the self-important belief that development represents the solution to the problems of the Third World, rarely considering the possibility that it might in fact be contributing to these problems, and that the development industry is constantly reinventing itself as the remedy for the ills that it causes (Crush 1995: 16).

Academics who occasionally deign to get away from their ivory towers are accused of participating in 'rural development tourism' (Edwards 1989: 122) and a kind of 'intellectual voyeurism' that makes not an iota of difference to the development problems that they purport to engage. In Edwards' opinion, this is anathema to the proper role of development research, which he claims should be contributing to the lives and promoting the development of poor and powerless people around the world (Edwards 1994: 281).

Clearly Edwards, as a development practitioner, and both Booth and Corbridge, as development scholars, are poles apart in their identification of the ultimate role and contribution of development research, and also the means of achieving this, although it is fair to point out that Edwards softened his position somewhat between his 1989 and 1994 articles. Echoing Corbridge's earlier-mentioned comment about how to 'change the world', Edwards (1989: 124–125) claims that:

> We cannot change the world successfully unless we understand the way it works; but neither can we understand the world fully unless we are involved in some way with the processes that change it. Development studies today are divorced completely from these practical processes of change.

In order to strengthen our understanding of the development process, and to enable us to make a more meaningful contribution to the lives of the world's poor and disadvantaged, a much more direct involvement in the process of development itself is required (*ibid*.: 125) and, echoing Brohman, a greater disposition towards 'listening and learning "from below"' (*ibid*.: 127) and an 'eclectic' approach that generalises *from* the particular (Edwards 1994: 286), that synthesises into a theoretical framework the diverse everyday experiences of ordinary citizens (*ibid*.: 287). Only by constructing theory from real experience can it have real explanatory power. Participatory research represents the best means of assuring relevance because the research 'grows out of people's own perceptions and is controlled by them' (Edwards 1989: 128).

One problem with the above sentiments is that the field of development has become rather too compartmentalised to be able to yield developmentalists who are capable of making the 'best of both worlds' – that is, saying something genuinely meaningful about development on the basis of diverse contextualised realities. In Edwards' own words (1994: 286):

> Practitioners in the past have focused too much on site-specific detail and have ignored the importance of wider economic and political frameworks [whereas] academics have focused too much on wider

forces in their search for grand theory, while neglecting the infinite variation of the real world.

Brohman (1996: 326) concurs: 'theoreticians and practitioners occupy different worlds and speak different languages [leaving] a yawning gap between theory and practice.'[3] It is my contention that the academic area specialist holds some potential to help to bridge this gap. The work of the area specialist should be sufficiently grounded and nuanced to be able to grasp local contextual realities, and yet this person must be sufficiently removed from 'extreme localisation' to be able to see how a clustering of 'trees' comes together to form a development 'wood', and how these, in turn, come together to form a definable 'ecosystem'.

The third line of argument concerns what Terry McGee (1995: 205) calls the 'vice of Eurocentrism' in both development theory and practice, and a 'world order created by Eurocentric images of the world' (*ibid*.: 192). The 'new development' advocates challenging the hegemonic privileging of Western knowledge in the field of development by asserting 'the value of alternative experiences and ways of knowing' (Crush: 1995: 4). The call is for intellectual emancipation via the 'indigenisation' of development – building development theory and praxis around indigenous concepts and practices: '[t]he indigenization of development thinking has become a central element of attempts to create more comprehensive and relevant approaches' (Brohman 1996: 337). In its purest form, indigenisation should centre around the privileging of indigenes (although note the caveat in the Preamble, and see below), but a watered-down version might also involve Western development scholars paying greater attention to indigenous knowledge systems.

Terry McGee is one among a growing number of development scholars who have attempted this. He charts a process of evolution in his own distinguished career whereby he progressed from being a scholar who was preoccupied with the theoretical ordering and interpretation of urbanisation processes in Southeast Asia, and the powerful influence of global forces on national and sub-national development, to his dawning realisation of the inherent weaknesses of models that derived from the Western experience (McGee 1995: 200). This led to a growing appreciation, via the global–local dialectic, of the way in which local conditions were interacting with national and international processes in both space and time, leading him to adopt 'a more organic, Asia-centred approach' (*ibid*.: 205). His concept of the extended metropolitan region, arguably a manifestation of a uniquely Pacific-Asian form of urbanisation which is heavily influenced by prevailing historical, ecological and cultural contexts, is presented as an approach to the study of development that attempts to

---

3. Such a dualistic framework conveniently eliminates the possibility that practitioners, while purporting to be working at the grassroots level and adhering to the local participation doctrine, may in reality be weighed down by a considerable amount of historical, cultural and experiential baggage that prevents their pursuing participatory research in its purest form. Similarly, there is an assumption that a theory-inclined development scholar is incapable of thinking globally and locally simultaneously or, in relation to the following discussion, of simultaneously being a development theorist and an area specialist – although this is quite possible within my own working definition of an 'area specialist' (see below).

privilege the indigenous over a hegemonic Western conception of reality. Areal specialisation would appear to have been quite an important factor in enabling such an epistemological transformation to take place. McGee (*ibid.*: 199) claims that geographers (and those of other disciplines) can only hope to foster indigenised theories of development 'if they have the right kind of cultural sensitivity'.

Crush (1995: 19) is nonetheless sceptical of the ability of Western scholars to 'formulate alternative ways of thinking and writing' to the extent that they can sufficiently 'indigenise' their research. McGee, too, remains sceptical about the possibility of indigenisation becoming the dominant script of development studies, but for institutional rather than intellectual reasons. The mainstream study of development, McGee claims, is so heavily constrained by both disciplinary boundaries and disciplinary orthodoxy that the prospects for a fundamental reorientation of development thinking and practice are extremely limited. Brohman (1996: 325) supports this view:

> discipline-centrism is an ongoing problem in development studies. The development process is artificially fragmented and compartmentalized to fit the areas of specialization, research methods, and theoretical frameworks of individual specializations. [In contrast] interdisciplinary approaches to development have yet to gain much respectability in an intellectual environment which tends to favor more 'scientific' and 'rigorous' research in disciplinary specializations.

As the following discussion will argue, areal specialisation – most particularly that which occurs without respect for disciplinary boundedness – represents a potential means of surmounting this 'impasse', and in the process making a valuable contribution to the 'new development'. This is because a deep comprehension of contextuality and the 'grounding' of development theory are argued to be essential to appropriate development thinking and effective development action.

Area specialisation

On the basis of the foregoing discussion, the 'ideal' development scholar would appear to require many if not all of the following qualities (and presumably several more besides). He/she would be well informed in current development debates, and also active in contributing to these debates. He/she would be very active in the field of development research, and this would involve a considerable degree of hands-on, grassroots-level fieldwork, conducted with and through local individuals and institutions (although note from the Preamble some of the difficulties that may be associated with this). If this person is not from the 'target' society, he/she should be fully versed in the precise workings of the local context – something that can presumably be obtained only after a considerable degree of contextual immersion: literature-based learning in this regard is not sufficient. This ideal person should have adequate skills of language and cultural understanding, should have the simultaneous powers of detachment and involvement, and should pursue a holistic understanding of the problems and process of development. He/she should be flexible, self-aware, humble, altruistic, sensitive, compassionate and a good listener, and should seek to contribute knowledge as well as acquiring it. This person should also, somehow, seek to achieve an appropriate balance between breadth and specialisation, focus and comparison.

The point I wish to make here is that several of the above qualities can best be nurtured through areal specialisation. What does this entail? There are obviously no hard and fast rules, but the following might be considered broadly indicative:

- *Experience*:   Clearly, the degree to which one becomes familiar with and aware of a particular context is partly a function of the amount of time one invests in contextual immersion. Many of the subtle aspects of life and livelihood in the developing world, and the complex cultural, political, social, historical, etc. structures that surround them, take time to understand, and it is common for one's understanding to change quite significantly over time. It is unusual indeed for someone to be able to acquire a deep and profound understanding of the host society simply by flitting in and out, how ever frequently this may occur. It is also the case that the contextual environment quite typically changes over time too, and thus a long-term and regular association with a locale spread over several visits may be as important as a single extended period of fieldwork: how many of us are considered 'experts' on areas which we have not visited for years? However, time alone is not a sufficient factor. The circumstances surrounding one's interaction with the 'target' society are just as important. Some people may spend years in a research locale without ever attempting or being able to understand it. Thus other factors are also important, especially the following two categories.

- *Communication*:   Effective communication is an important key to understanding a host society, and this ideally requires that (where necessary) considerable skills are developed in the language(s) of the host society (as well as non-verbal means of communication). Information can be obtained through an intermediary, but I do not think that this is generally adequate for the purposes of penetrating beneath the surface of culture, attitude and context. Communication is not just about language but also about the nature of the relationship between two parties, and also about vocabulary that is used and the way that answers and opinions are heard and interpreted.

- *Personality*:   Ultimately, I believe that the personal dimension is a key variable. A myriad facets of personality might be discussed, but I shall touch on just a handful. Of paramount importance is self-awareness. One has to understand who one is, and where one is coming from in terms of culture, background, experience, training, philosophy and so on if one is to be able to shed certain forms of 'baggage' in order to 'see' or 'experience' things in the way that a target society does – which I believe is a key to effective understanding and thus action. An ability to recognise, for instance, Eurocentrism, stereotype, romanticism and so on is crucial if we are to minimise the influence of these things on our judgement, perceptions and action recommendations. Linked to this is what I call 'immersibility' – the willingness and ability of the researcher to become deeply imbued within the context of the problem or phenomenon with which he/she is concerned, as opposed to retaining a degree of detachment and thus isolation. The researcher must possess skills of both intuitive and objective interpretation of events, opinions, actions, processes, circumstances and settings, and also the ability to recognise appropriate signs that might warrant interpretation. Flexibility is also important. Two further traits of personality that I personally believe are indispensable are altruism and humility (see also Edwards 1989). On this basis, it may be perfectly possible for someone to be working at the grassroots level but failing to achieve the requisite degree of contextual penetration, immersion and understanding if they do not possess the appropriate qualities of personality, attitude and ideology.

- *Meaning of development*:   We should not overlook how scholars interpret the meaning of 'development', and also how these interpretations are derived, as a further factor that might influence the effectiveness of their contribution 'where it counts'. I would argue that there is a significant difference between a case where someone learns about 'development' from a textbook and then seeks to apply one's knowledge to developing world realities, and someone whose interpretation of the meaning and requirements of development is informed by deep and profound experiences of the developing world. This roughly parallels the earlier discussion of the 'top-down' and 'bottom-up' pathways to the generation of theory and the engendering of understanding.

- *The reality of development*:   It does not necessarily follow that the development researcher who possesses all of the above qualities will be able to make a meaningful contribution to the development experience of others in the manner that Edwards (1989, 1994), in particular, advocates. The cases presented in the Preamble show very clearly how a great many factors that make up development realities may present themselves as formidable barriers to effective development action, irrespective of how well versed the development researcher is in contextuality and locality. Career interests, corruption and, most ironically, the rigid (often self-seeking) adherence to a Western development orthodoxy by the upper echelons of power hierarchies constitute just some among many of the development realities that lie in the path of altruistic developmentalism.

It would therefore appear that an important criterion in areal specialisation is what Benny Farmer (1973: 9–10) has called 'dedication to area'. It is generally the case that most (but far from all) academics who work in the field of development will, at some stage in their careers (most particularly when undertaking their PhD), have undertaken a period of field-based research in a particular developing area (region, nation, sub-nation; individually or comparatively). However, while this process may also have involved the 'ground-truthing' of development theory, it does not follow that such researchers will become equally dedicated to, specialised in, or contextually immersed within this developing area or areas. This may be explained partly by opportunity and partly by inclination.

The academic study of development is principally located under three, or perhaps four, umbrellas: disciplinary departments, development studies institutions, area studies institutions, and beyond academia. Bruce Koppel (1995: 4–5) draws a distinction between the first three of these in terms of the extent to which they both encourage and allow scholars to build 'understanding based on the holistic analysis of a specific place or culture'. Within area studies, locality, contextuality and multidisciplinarity are pre-eminent, and a strong emphasis is placed upon developing a facility with the local language(s) and upon building extensive field experience (*ibid.*: 4). Development studies also encourages a holistic analysis of development processes and problems, although the developing area or region becomes less important as the basis for the accumulation of knowledge, even though it may represent an 'arena' where theoretical and methodological issues are tested out (*ibid.*: 4). Language and 'hands-on' field experience are less heavily emphasised. In academic disciplines, by contrast, the locality has generally been far less important than the refinement of theory and method (*ibid.*: 5), and the study of development becomes fragmented by disciplinary boundaries, at a cost to holistic understanding. The possession of linguistic and cultural skills is not an absolute requirement.

Based on this highly simplified typology, it can be seen that the study of development according to the criteria which have emerged from its recent rethinking would appear to be much more difficult within disciplinary departments than it is outside them.[4] However, it is generally the case that area studies, and to a lesser extent development studies, are 'hostage to the hegemonic grip of disciplines' (Hirschman, *et al.* 1992: 4) that claim for themselves a certain primacy of status and method. Bruce Koppel (1995) claims that the exponents of area studies are quite typically seen as 'refugees' from mainstream disciplines,[5] reflecting a distinct lack of status and, by extension, an extreme political vulnerability. At the same

[4.] This is obviously a gross oversimplification of the situation. In the United Kingdom, the very active membership of the Developing Area Research Group (DARG) of the Royal Geographical Society/Institute of British Geographers, consisting of some very capable 'area specialists', clearly demonstrates that Bruce Koppel's typology (which relates to the situation in the United States) does not apply equally in all cases. Nonetheless, it would be interesting to conduct a straw poll among the membership of DARG to ascertain the extent to which members feel either central or marginal to their departments and disciplines, and upon what basis their centrality (perhaps their engagement of meta- and top-down theoretical formulations) or marginality (perhaps areal specialisation and 'dirty feet') rests.

[5.] Although Koppel discusses the situation in the United States, I believe that his sentiments are more generally relevant. As an academic who has long been associated with both a disciplinary and an area studies department, I have been aware on countless occasions of advice offered to young and aspiring academics from senior figures within geography that area studies represents a career dead-end; a graveyard full of empiricists, esotericists and ephemeralists. As a riposte, I have written elsewhere (Parnwell 1996) of the excellent geography that in the past has been written by such Southeast Asia specialists as Charles Fisher, Jim Jackson, Karl Pelzer, Pierre Gourou, Charles Robequain and Joseh Spencer that, by and large, has remained largely invisible to the mainstream of the discipline, as also tends to be the case with current Southeast Asian geographers.

time, Koppel asks whether area specialists might instead claim to be 'settlers' within a new academic terrain, and I believe that the argument presented in this chapter would support this contention in principle, even though it is unlikely to gain much ground in practice. As such, it might be argued that the position of area studies within academe is weak at precisely the time that area specialists are excellently placed to make a valuable contribution to the new development. Meanwhile, mainstream academics who are becoming interested in processes or theories that draw in peripheral societies, such as the changing international division of labour, or globalisation, increasingly 'dabble' in these societies without ever threatening to understand them.

Conclusion

This chapter is not intended as a party political manifesto on behalf of area studies; nor does it seek to make territorial claims around particular localities or epistemologies. Rather, it seeks to encourage better development scholarship by drawing attention both to the importance of contextuality in the 'new development' and to the contribution that the area specialist can make in weaving this into a broader understanding of the development process. As such, it claims that the academic study of development should be based around a continuum, where theory and reality are complexly interwoven, rather than a dichotomy, where they occupy essentially different worlds. By building a conceptual understanding of the development process from the bottom up based on a deep understanding of contextual realities, a way out of the impasse in development studies may be found that allows us to rise to the challenge of 'generalising and theorising diversity' without creating a 'mish-mash of ideas and approaches' and 'an anarchic lack of coherence' (Booth 1994).

This chapter offers a muted challenge to orthodoxy by suggesting how the conventional structuring of development research within and outside disciplinary departments, and the power relations and rivalries that accompany this system, represents a significant impediment to the advancement of the 'new development'. Ultimately, however, the issue boils down to a question not so much of *where* development scholarship is undertaken but *how*. Just as 'dedication to area' has been argued to be a virtue in the new development, so 'over-dedication to area' may paradoxically be considered a vice. On the other hand, recent developments in cultural studies, through which disciplinary departments (in the humanities) have become concerned with contested representations of the other, are to be welcomed in the present context because, rather than representing a further challenge to area studies for leadership in international scholarship, as Bruce Koppel (1995: 17) warns, it may be interpreted as one of several signs that the academic mainstream is taking seriously, and responding to, the critique of orthodox scholarship that the 'new development' represents and requires.

References

Booth, D. (ed.) (1994) *Rethinking Social Development: Theory, Research and Practice*. Longman, London.

Brohman, J. (1996) *Popular Development: Rethinking the Theory and Practice of Development*. Blackwell, London.

Corbridge, S. (1993) Marxisms, modernities, and moralities: development praxis and the claims of distant strangers, *Environment and Planning D: Society and Space* **11**: 449–472.

Corbridge, S. (1994) Post-Marxism and post-colonialism: the needs and rights of distant strangers. In Booth, D. (ed.) *Rethinking Social Development: Theory, Research and Practice*, Longman, London, 90–117.

Cowen, M.P. and **Shenton, D.W**. (1996) *Doctrines of Development*. London: Routledge.

Crush, J. (ed.) (1995) *Power of Development*. Routledge, London.

Edwards, M. (1989) The irrelevance of development studies, *Third World Quarterly* **11**(1): 116–135.

Edwards, M. (1994) Rethinking social development: the search for 'relevance'. In Booth, D. (ed.) *Rethinking Social Development: Theory, Research and Practice*, Longman, London, 279–297.

Escobar, A. (1995) *Encountering Development: The Making and Unmaking of the Third World*. Princeton University Press, Princeton, NJ.

Farmer, B. (1973) Geography, areas studies and the study of area, *Transactions of the Institute of British Geographers*, 60: 1–16.

Galli, R.E., Rudebeck, L., Moseley, K.P., Stirton Weaver, F. and **Bloom L.** (1992) *Rethinking the Third World: Contributions Toward a New Conceptualization*. Crane Russak, New York.

Ginsburg, N., Koppel, B. and **McGee, T.G.** (eds) (1991) *The Extended Metropolis: Settlement Transition in Asia*. University of Hawaii Press, Honolulu.

Halib, M. and **Huxley, T.** (1996) *An Introduction to Southeast Asian Studies*. Tauris Academic Studies, London.

Hirschman, C., Keyes, C.F. and **Hutterer, K.** (eds) (1992) *Southeast Asian Studies in the Balance: Reflections from America*. The Association for Asian Studies, Ann Arbor, Michigan.

Koppel, B.M. (1995) *Refugees or Settlers? Area Studies, Development Studies. and the Future of Asian Studies*, East–West Center Occasional Papers (Education and Training Series) No. 1 (April), East–West Center, Honolulu.

Lehmann, D. (1997) An opportunity lost: Escobar's deconstruction of development, *Journal of Development Studies*, **33**(4): 568–578.

McGee, T.G. (1995) Eurocentrism and geography: reflections on Asian urbanization. In Crush, J. (ed.) *Power of Development*, Routledge, London, 192–207.

McGee, T.G. and **Greenberg, C.** (1992) The emergence of extended metropolitan regions in ASEAN: towards the year 2000, *ASEAN Economic Bulletin.* **9**(1): 22–44.

Minogue, M. (1988) Problems of theory and practice in development studies. In Leeson, P.F. and Minogue, M.M. (eds) *Perspectives on Development: Cross-Disciplinary Themes in Development Studies*, 224–250.

Mittelman, J.H. (1995) Rethinking the international division of labour in the context of globalization, *Third World Quarterly* **16**(2): 273–295.

Moore, D.B. and **Schmitz, G.J.** (1995) *Debating Development Discourses: Institutional and Popular Perspectives.* Macmillan, London.

Parnwell, M.J.G. (1996) Geography. In Halib, M. and Huxley, T. (eds) *An Introduction to Southeast Asian Studies*, I.B. Tauris, London, 101–147.

Parnwell, M.J.G. and **Khamanarong, S.** (1990) Rural industrialisation and development planning in Thailand, *Southeast Asian Journal of Social Science* **18**(2): 1–28.

Parnwell, M.J.G. and **Khamanarong, S.** (1996) Rural industrialisation in Thailand: village industries as a potential basis for rural development in the Northeast. In Parnwell, M.J.G. (ed.) *Uneven Development in Thailand*. Avebury, Aldershot.

Parnwell, M.J.G. and **Wongsuphasawat, L.** (1997) Between the global and the local: extended metropolitanisation and industrial location decision-making in Thailand, *Third World Planning Review* **19**(2): 119–138.

Porter, D., Allen, B. and **Thompson, G.** (1991) *Development in Practice: Paved With Good Intentions*. Routledge, London.

Schuurman, F. (ed.) Beyond the Impasse. *New Directions in Development Theory*. Zed Books, London.

Trainer, T. (1989) *Developed to Death: Rethinking Third World Development*. Green Print, London.

CHAPTER 5

Contextualising professional interaction in Anglo-(American) African(ist) geographies

Reginald Cline-Cole

> [G]eographers have tended to focus their analysis of the spatiality of everyday life on the social relations **external** to geography – the object of 'our' gaze has thus tended to be **out there**, in **the field**. ... To date, there has been little attention paid to the way in which 'we' as geographers discursively constitute the 'conceptual space' ... of geography through ... **banal** disciplinary practices.
>
> (Berg and Kearns 1998: 129; emphasis in original)

As geographers increasingly turn their gaze 'inwards', both their discipline and the social relations implicated in its production become important 'fields' of analytical reflection. Resultant change in geographical sensibilities has, among other consequences, stimulated new interest in the problematics of research on and in, and the teaching of developing countries by Anglo-American geographers (Corbridge 1993; Harrison 1995; Kay 1993; Lonsdale 1986; Potter 1993; Rundstrom and Kenzer 1989).[1] This revival includes a (re)consideration of the need and motivation for professional interaction between academic geographers of the North and South. It also addresses the nature, extent and consequences of such interaction. Overall, its preoccupations are a logical extension of current concern with questions of professional ethics, including responsibility to distant others or strangers (Harrison 1995; Madge 1993; Porter 1995; Proctor 1998; Sidaway 1992, 1993,1997a; Smith 1994).

Consequently, there have been recent calls by some Anglo-American

---

[1.] I use the term 'Anglo-American' throughout this chapter in the restricted sense of geography centred in Britain and America, without implying homogeneity in the separate geographic traditions that the description encompasses, and while being mindful of its potential for exclusion (*cf.* Berg and Kearns 1998). With reference to the latter observation, it is worth noting that the discussion in this chapter excludes mainland European geographic traditions, which are also implicated in the (re)making of academic geography in (Anglophone) Africa, but which require detailed treatment in their own right. I use the term 'Anglo-American' interchangeably with 'Northern', which I contrast with 'Southern' or 'Anglophone African'. I also use the terms 'First World' and 'Third World' and 'developed' and 'developing' countries, in keeping with the practice of many of the authors I cite. Needless to say I am aware of the need to maintain a critical distance in the use of these labels, which, as binary distinctions, minimise diversity, complexity and heterogeneity.

geographers for greater but less exploitative interaction within the world community of professional geography (Paul 1993; Smith 1994). Notably, during his presidency of the Association of American Geographers (AAG), Tom Wilbanks evoked the imperative for internalising such unimpeachable professional ethics in interaction between the better-endowed and less fortunate sectors of this global disciplinary community. In essence, this 'new' thinking about the practice of interaction proposes greater resource sharing as a response to uneven global academic development. It envisages that such exchange will take place within the wider context of a globalising project of disciplinary and institutional incorporation, and that such a project will benefit all geographers (Paul 1993; Wilbanks 1993).[2]

Such proclaimed inclusiveness notwithstanding, new interaction is conceived as a project to be directed by Anglo-American geography. Thus American geographers 'can start by getting better informed, and ... can start to show leadership in sharing ... fairly meagre resources' (*ibid*.: 2). Similarly, British geographers 'should at least consider the possibility of contributing to a wider responsibility to the potential "world community" of professional geography' (Smith 1994: 366). Significantly, however, new interaction strenuously repudiates all imperialist intent. Thus Porter (1995), for instance, raises the possibility of a key role for African and other developing country nationals employed in the academy in the North. She encourages them to 'spearhead ... new, more equitable, forms of collaboration, on the basis of their particular knowledge of local conditions in the countries concerned' (Porter 1995: 141).

Despite its non-prescriptive nature, this paper is a direct response to Porter's implied challenge. It contains the reflections of an ambivalent African academic geographer, currently employed in a British university, who has 'played' both the Northern and Southern ends of the 'interaction game', and who has also had first-hand experience of the conditions under which academic geography is produced in both the North and the South.[3] Its overall aim is to map the territory on which ideas about 'new' professional interaction have evolved during the 1980s and 1990s. It does so with reference to Anglophone African and Anglo-American geographers and geographies. However, 'geography is a social and historical discourse, which is always intimately bound up with questions of politics and ideology' (ÓTuathail and Agnew 1992: 190). Thus any intensification in current practice must be based on a finely textured understanding of the political, economic and social circumstances underlying such exchange. Similarly, the history of recent practice also contains valuable

---

[2.] Wilbanks wrote about the process of incorporating or 'mainstreaming' change in 'traditionally' uneven or exploitative relationships in collaborative North–South research arrangements. This chapter extends this restricted meaning to cover the idea of a 'new realism' or 'new deal' in *all* aspects of academic exchange or interaction between Northern Anglophone and Anglophone African geographers and geographies. I use the description 'new interaction' (thinking and practice) throughout as shorthand to refer to this, while remaining aware of the risks of such 'labelling' or 'naming'.

[3.] These reflections are the product of a personal career trajectory whose contours have been defined largely by the historically dominant forms of North–South interaction in Anglo-(American) African(ist) geography, which form the subject of this chapter.

lessons for any future transformation. I attempt therefore to provide such a pluralistic context to the production of ('new') interaction thinking and practice.

More specifically, I have three objectives: (1) to highlight the role of material transformations affecting Anglo-American and African geographies, which appear to be challenging contemporary sensibilities (*cf.* Dodds 1994; ÓTuathail 1992; Sidaway 1997a; Smith 1994); (2) to show how varied and uneven experiences of these transformations influence the philosophy and practice of academic exchange; and (3) to argue for politics to be considered a core element of all projects of professional interaction, given that these are ultimately about the macro- and micro-politics of access to and control over human, intellectual and material (academic) resources (*cf.* Leftwich 1983). My intention here is to emphasise the unity between the politics, economics, psychology and 'morality' of professional interaction, and to show how this unity manifests itself in both thinking and practice. The aim is to reintegrate politics into debates that increasingly treat it as peripheral because they appear to 'accept' the social institutions of Northern and Southern geographies as they exist (ÓTuathail and Agnew 1992).

I do not, therefore, present a review of interaction between Anglo-American academics and African scholars *per se*. Rather, I attempt to identify some of the wider questions raised by an examination of interaction in context. I focus on the latter's inherent ambiguities and contradictions. Through this, I hope to highlight the ethical and practical dilemmas that professional interaction poses for geographers. The chapter is organised into six parts, including this introduction. The next section attempts to situate new interaction thinking within currents of disciplinary and personal introspection. The third and fourth sections address, respectively, the changing contexts and conditions under which Africanist and African academic geography is produced. The fifth section then locates practices of professional interaction and the 'new' thinking about the nature/philosophy of such interaction within this wider context of transformation/change. The complex and frequently contradictory dynamic that emerges necessitates some examination of the variety of individual/group/institutional experiences of interaction. This section thus concludes with brief examples of how these experiences are retold and redeployed in constructing 'meanings' around the notion of professional exchange. I identify some of the ethical and practical dilemmas that the foregoing poses for North–South professional interaction in the sixth and concluding section, which also reiterates the crucial (albeit circumscribed) role of personal responsibility in the forging of a new ethos of professional exchange.

Problematising the 'new' interaction

Recent calls within Anglo-American geography for 'reaching out to the rest of the world' (Wilbanks 1993: 2) or assuming 'professional responsibility to distant others' (Smith 1994: 359) derive to varying degrees from three interdependent areas of recent academic concern among geographers. First, there is a reported decline in interaction between geography in the North and in the South (Dickenson 1986; Lonsdale 1986; Paul 1993; Robinson and Long 1989; Unwin and Potter 1992). Second, there is increasing concern over the implications of such reduced interaction for the progress of geography in developing areas (Gilbert 1987; Paul 1993; Porteous 1986; Potter 1993; Shresta and Davis 1989), and for the more general reinforcement of Eurocentricity/ethnocentricity in Anglo-American geographic discourses (Bradshaw 1990; Johnston 1984; 1985; Rogerson and Parnell 1989; Sidaway 1990; 1997b; Slater 1989; Thrift 1985). And, third, there is growing appreciation of the very limited extent of critical awareness of the full range of redistributive consequences associated with the practice of professional interaction/exchange (Crush *et al.* 1982; Madge 1993; Rogerson and Parnell 1989; Sidaway 1992).

At the same time the notions of a 'new' interaction to which such concerns give rise do not represent an undifferentiated or even particularly coherent body of thought, suggestions, beliefs or practices. For example, in addition to a firmly held conviction that greater professional interaction would benefit all geographers through closer association of geographies worldwide, Bimal Paul also believes that intensified interaction will promote personal and professional discovery; narrow the 'technical and conceptual gap [separating] the geography of the two worlds'; and stimulate debate about resource sharing among geographers (Paul 1993). In the process he appears to demonstrate, maybe not entirely intentionally, a tendency to underplay differentiation within Anglo-American and African geographies as well as between geographers in these two traditions. And, arising from this, to underestimate the significance of the unevenness that is likely to characterise the production of a 'new' interaction. This is tantamount to divorcing academic interaction from the wider socio-economic, political and cultural context that gives it meaning.

In contrast, David Smith offers a more circumspect assessment. While noting the potential of certain kinds of interaction for confronting and ultimately undermining patterns of academic inequality and uneven development, he cautions against the unwitting use of uncritical academic exchange in reproducing and reinforcing existing relations of domination and oppression (Smith 1994). Politics and social histories are integral to his explanations of why Anglo-American geographers can and should do more to develop more generous institutional ethics. Indeed, he privileges notions of (im)partiality, domination and oppression, in ways which suggest that uneven academic development should not be accepted simply as a 'given', with interaction setting out merely to mitigate its worst impacts.

For him, the 'new' interaction should be an integral part of a wider assault on the ultimate causes of uneven academic development. He is worth quoting at length:

[W]e (collectively in the advanced capitalist world) bear some responsibility [for the position of disadvantage of others]: uneven academic development is in part an outcome of international structures of domination and exploitation. ... Like place of birth, finding oneself a member of staff of a geography department in Britain as opposed to, say the Republic of [Kenya], carries no moral merit; it is merely a matter of personal good or bad fortune. It is therefore impossible to justify such gross discrepancies with respect to the resources for scholarship which exist among such places, except as a reflection of position in broader hierarchies of economic and political power which themselves have no moral justification.

(ibid.: 364)

His is a vision of interaction that is reformative in ways Wilbanks (see below) merely hints at. It is also a vision that is subversive in a manner that both Wilbanks and Paul diplomatically avoid altogether. At the same time, although he remains realistic about the limited prospects of this vision, he refuses to accept this as enough reason to capitulate too easily to the advance of the forces of neoliberalism, and their promotion of the commodification of knowledge production and exchange (see also Madge 1993; Mohan 1994). Thus his understanding of 'professional geographical responsibility' is in some ways similar to, but in several important respects quite different from, that of both Paul and Wilbanks.

Treading a fine line between Paul and Smith, Tom Wilbanks (1993) hints at an appreciation of the reasons behind Smith's caution. He is adamant, for instance, that academic colonialism should have no place in professional interaction, and he favours a policy of withholding funds for overseas research from Anglo-American geographers whenever their research proposals fail to make allowance for meaningful host country participation. But he also makes a spirited argument for greater North–South interaction that is reminiscent of both Smith and Paul. Like the latter's, however, his is a vision that is largely devoid of controversy. Thus he appears to suggest, probably unintentionally, that the motivation behind recent calls for greater professional interaction might be of less immediate importance than a recognition of the pressing need for the intensification of such exchange in the first place (*ibid.*). If this is indeed so, it would be minimising the true significance of the myriad influences that shape individual and group motivation. It would also be to neglect the role of such motivation in structuring the aims, goals and nature of interaction. Yet professional interaction conjures up profoundly varied 'meanings' for individuals/groups/institutions (*cf.* for instance, Batterbury 1997 and Sidaway 1992).

Overall, Wilbanks recognises that interaction is a dynamic and uneven process subject to continuous (re)negotiation. He offers, as do both Smith and Paul, practical suggestions for negotiating a proposed 'new' interaction. However, unlike Smith, but like Paul, he does not address the politics of interaction directly. For instance, although he refers in passing to a history of North–South exchange between geographers and geographies, he fails explicitly to engage questions of the complex and contradictory ways in which such a 'past' is likely to be remembered in both North and South. Equally significantly, he fails, again like Paul, to anticipate contestations of his vision, particularly from geographers and other academics in the South. And yet, even in the midst of the widespread professional 'hunger for interaction with the international scholarly community' that he senses among Southern colleagues, many of the latter might conceivably be suspicious of his true motives (and those of Northern colleagues in general) in 'reaching out' to them in the manner envisaged. Histories of interaction are not only experienced, remembered and retold in different ways; they are also redeployed in a variety of practices in the production of new geographies and sociologies of interaction.

Inevitably, 'new' interaction thinking is premised on particular readings of the material, cultural, moral and social geographies of North–South academic exchange. For as ÓTuathail and Agnew observe, 'the way we describe the world, the words we use, shape how we see the world and how we decide to act' (1992: 190). All three authors cited above evoke, admittedly in different ways, images of Anglo-American (moral and intellectual) leadership; African (academic and material) hunger, helplessness and decay; and (yawning) conceptual and technical gaps separating Northern and Southern Anglophone geographers, which need to be bridged. They identify compelling professional, personal, moral and practical reasons why Anglo-American geography and geographers could and should be less Eurocentric and ethnocentric. They also suggest means, modalities and agency for the realisation of their shared vision (see below).

Yet such theorisations constitute no more than particular contributions to the (re)production and modification of competing sensibilities or interaction 'ideologies' (*ibid.*). For example, one of Wilbanks' recipes for greater professional interaction is, rather paternalistically, for American geographers to 'adopt' overseas geography departments. And, as with all ideologies or representations, those cited here are suffused with biases. Furthermore, such biases are partly informed by a relative Anglo-American ignorance of other geographies and geographers (Paul 1993; Smith 1994; Wilbanks 1993). These raise important questions concerning the disciplinary and institutional contexts within which such theorisations are produced.

African(ist) 'other(s)' in Anglo-American geography

Producing academic geography

The conditions under which academic geography is produced, and the status of the discipline, are continuously renegotiated and redefined. Within the British academic community, current transformations have origins that can be traced back to the years of deep and prolonged economic crisis, which succeeded an initial period of rapid postwar expansion in tertiary and geography education up to the late 1970s. They are occurring against a background of uneven public expenditure cuts, the increasing influence of higher education funding agencies, and 'growing state involvement in the structure and priorities of ... universities' (Smith 1986: 239). Not surprisingly, they have generated a diversity of disciplinary, institutional, group and personal experiences, 'as numbers of students and full-time staff fell for the first time since the end of the Second World War ... and the job market for new academics became much more limited in terms of the number of new posts' (Sidaway 1997a: 491). Reviews of the state of contemporary North American academic geography expressed similar concerns (Haigh 1982; Lonsdale 1986; Porteous 1986; Robinson and Long 1989; Rundstrom and Kenzer 1989).

The restructuring of Anglo-American academic geography has continued into the 1990s. Sidaway demonstrates how '[o]ne of the consequences is a general tendency towards intensification – within a context of increasing unevenness – in the management and work practices of institutions/ departments and of individual roles within them' (Sidaway 1997a: 492). Against such a background and, given that the labour market for geographers in the United States has shrunk quite substantially during the 1990s, many – maybe most – Anglo-American academic geographers remain preoccupied with individual, group and disciplinary survival and growth, despite the continuing popularity of their subject among students (Batty 1994; Wilbanks 1993).

Contesting place

Geography's continuing popularity among students frequently obscures the changing countours of both the division of labour within the discipline as a whole and the balance of power/influence between its various sections. Within this wider context, Anglo-American geography 'described, classified and ordered [its] various region[al specialisations] and [their practitioners] through the construction of highly arbitrary imaginative geographies' (Dodds 1994: 96). In the process, the role of African(ist) and other 'Third World' geography was redefined and the place of its practitioners renegotiated (Barbour 1982; Farmer 1983; Potter 1993). This needs to be considered as further evidence of the uneven impact of the restructuring of the discipline discussed in Sidaway (1997a). As a result, and despite the continuous growth in the geography student population in higher education, British-based developing areas' geographies apparently fared especially poorly (Dawson and Hebden 1984; Farmer 1983; Lawler 1992). It is therefore worth exploring the dynamics and outcomes of the processes of transformation set in motion.

Gilbert, for instance, felt obliged to decry the fact that 'the study of the Second and Third Worlds is regarded as being separate from, and by implication inferior to, mainstream [British] geography' (1987: 145). That his intervention came almost four decades after Wooldridge's notorious assessment in an Institute of British Geographers' presidential address that 'the human geography of Somerset is more interesting and in many ways more significant than that of, say, Somaliland', is in itself richly suggestive (Wooldridge 1950: 7,9). Clearly, Wooldridge's caution about the 'real peril to the well-being of geography' posed by the 'disease of "otherwheritis"', resonated in some sectors of academia in a Thatcherite Britain (*ibid.*: 9). 'Thus,' as Berg and Kearns (1998: 129) observe in a related context, 'are geographies of other people and places ... marked as Other –

exotic, transgressive, extraordinary, and by no means representative.' The consequences were profound. Particularly badly

> hit [was] the provision of posts and the recruitment of young geographers to them; [and] research and travel funds for academics, would-be research students and would-be organizers of overseas research projects [which] grew ever more exiguous, as in the hungry thirties.
>
> (Farmer 1983: 77)[4]

Consequently, there was a relative decline in the number of First World academics and research students specialising in developing areas research and teaching, as well as an erosion in the status and funding of overseas (field) research, teaching and publishing (Gilbert 1987; Paul 1993; Potter 1993). Several older Africanists are convinced that their careers suffered as a direct result. Younger geographers worried about the implications for their employment, tenure and promotion prospects.

Such concern was no doubt fuelled by the additional consequences of intra-disciplinary (re)negotiation and restructuring. Geographers who teach primarily on the Third World had to operate in relative isolation in British university geography departments (Unwin and Potter 1992). Additionally, 'realistic' tuition fees for overseas students were introduced in the 1980s, when the image of these students as parasites living off British universities and, ultimately, taxpayers was invoked by the then Thatcher government in an attempt to win public support for the move. One commentator of the time observed how overseas student intake was discussed 'primarily as a source of revenue [with] admission turning on the ability to pay fees which have been inflated by creative official accountancy' (Hargreaves 1985: 330). In the event, these fees have proved prohibitive. British Africanist geographers and their departments continue to struggle to recruit overseas students, who have traditionally brought valuable diversity to the discipline (Farmer 1983). This loss of diversity is of more than passing significance. For although immigrants who retain a professional interest in their original homelands play an invaluable role in overseas research in North America, ethnic minorities are, overall, poorly represented within the community of Anglo-American geographers (Porteous 1986; Shresta and Davis 1989). The practitioners of British geography remain 'predominantly white, Anglo-Saxon and Celtic, and male' (Woods and White 1989: 201; see also McDowell 1990). To my knowledge, the discipline currently boasts fewer than ten

---

4. It should be noted that British Africanist geographies were suffering the same fate as African (and Area) Studies in general. Consider the following. 'Between 1977 and 1982 the proportion of the then Social Science Research Council budget spent on African Studies fell from 2.3 to 0.8 percent. During the "Thatcher cuts" of the early 1980s, travel grants at British universities were slashed, library book funds diminished, excellent scholars were forced into early retirement and some of the most lively departments in the country ... were axed and their members forced to move elsewhere. Under pressure to economize, universities transfered resources from area studies ... back into core disciplines where Africanists tended to be peripheralized. Funding for postgraduates dried up and vacant posts were frozen' (McCracken 1993: 243).

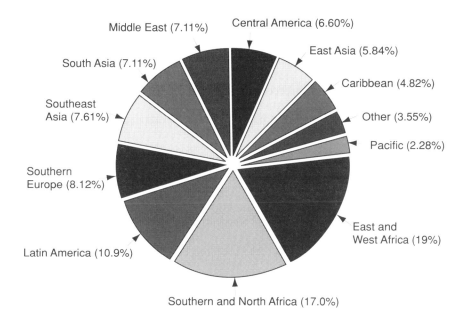

Figure 5.1   The teaching of developing areas' geographies in British
tertiary education (percentage of departments in which
developing regions are taught)
Source: graphed from data in Unwin and Potter (1992: 58)

African university lecturers, and, proportionally, a considerably smaller
representation of minorities overall than in North America.

Nonetheless, Third World/Developing Area courses remain popular
among undergraduates (Unwin and Potter 1992). Students are not just interested in learning about 'other places', they '[a]re keen to learn about
Third World issues and problems' (Harrison 1995: 286). Figure 5.1 suggests that geographies of Africa may well constitute a significant share of
this teaching 'market'. Increasingly, however, such teaching appears to be
taking place within general development courses, which continue to
grow in number, rather than in regional geography courses (Harrison
1995; Simon 1997). Harrison (1995) attributes this to the fact that regional
courses are limited in scope and diversity. Given the enormous potential
of this trend for the long-term peripheralisation of the area specialist
within mainstream geography, this clearly raises important questions
concerning power/knowledge within the discipline (Parnwell, Chapter 4
this volume). Would 'general' courses on the geography of the (global)
South or emerging market regions or developing areas, for instance, increasingly replace regional Africa courses (and, ultimately, the primary
teaching/research focus of the Africanists who teach them)? Can the
'turn' to development studies (and, increasingly, questions of globalisation)
be justifiably interpreted as one of the latest manifestations (however

unintentional) of prevailing prejudice against areal specialisation in British academic geography?[5]

What about British-based geographical research on Africa? Entries in the 1987 *Register of Current Research Interests of Members of the Developing Areas Research Group, Institute of British Geographers* suggest that Africa, the main focus of research, accounted for about 40 per cent of all research projects listed (Potter and Unwin 1988). The most recent issue of the register notes a 'continued predominance of the African continent as a focus of developing area research amongst U.K. geographers' (Parnwell 1994: i). The continent rated 25 per cent more entries than Asia, and almost as many entries as Latin America, the Middle East, and the Caribbean and Pacific Islands put together (*ibid.*). There is here, it seems, some confirmation of the place and role of Africa within research into developing areas in British geography reported in the late 1980s (Potter and Unwin 1988).

But the register points to the possibility of both imminent and more long-term change in these trends. It notes that although continued growth in developing areas research appears likely, this interest may not necessarily be UK-based or continue to favour Africa to the same extent. Thus some three-quarters of registered students researching developing area topics are themselves from developing countries, to which many are likely to return at the end of their studies (Parnwell 1994). Less commonly, such students are descendants of immigrants from these countries who, though they are almost certain to stay in the UK, are unlikely to end up working as academic geographers (see earlier comments regarding the social composition of British academic geography). In addition, research interest in Asia appears to be growing much more quickly than for any other developing region. Indeed, I know (former Africanist) geographers who have already (or are currently being) 'retrained' or 'respecialised' to take advantage of this expansion in research interest and associated funding in South Asia and the Pacific Rim. Similarly, I am aware that some people have been forced to change jobs because their developing area interests did not match newly defined research profiles of their old departments/colleges/universities.

Nonetheless, along with other disciplines in the British academy,

---

[5.] I ought to declare my position here. I am a geographer based in an African Studies centre. Yet this is not in any way to suggest that area studies is, or should be static/unchanging. Indeed, I agree that 'many important aspects of what affects the continent are external,' and that African studies 'should not be confined to the geographic entity we know as Africa' (Hyden 1996: 14). But this still leaves important issues unresolved. For instance, I know of a geography department in a British redbrick institution that has tried to justify a proposal to cancel the only regional Africa option available to its upper-level undergraduates on the ostensibly 'pedagogic' grounds that (1) the continent is referred to in thematic courses, and (2) other world regions like Eastern Europe and the former Soviet Union constitute more 'relevant' areas of study in a globalising world. Similarly, many developing area geographers in geography departments (rather than area studies units) currently find themselves under pressure to publish in 'mainstream' geography journals rather than area studies outlets. In other words, because disciplinary transformation can become a source of self-interest and parochialism (Smith, 1994), it is vital that these issues are debated within the discipline, as well as 'with those groups and institutions in society that have a bearing on our work' (Hyden 1996: 15). The threat of knowledge becoming decontextualised is one that must be taken seriously.

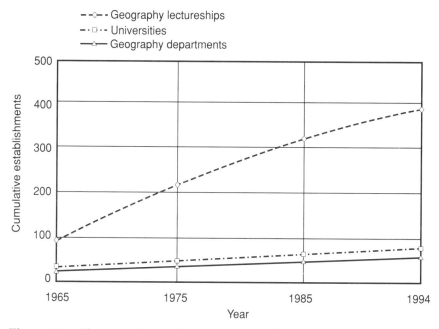

Figure 5.2 The growth of university geography in Commonwealth Africa (including South Africa). *Source:* Various issues of the *Commonwealth Universities Yearbook* and *World of Learning*.

(Africanist) geography 'should never forget how relatively limited is its resource crisis as compared with the much graver problems existing in African universities' (McCracken 1993: 242). Indeed, a number of historical similarities notwithstanding, the context within which British geography is produced is radically different from the conditions under which academic geography is produced in Africa. We examine these conditions next.

Geography in a deconstructing African academy

The nurturing years

Along with the rest of the continent's academic community, African geography experienced four decades of expansion, following the establishment of the first tertiary departments in the 1940s. From a base of fewer than thirty up to the 1960s, Anglophone Africa's universities grew to more than eighty by the mid-1990s. Together, these institutions contain a large number of geography departments, which offer a variety of disciplinary programmes and courses, and employ a steadily growing number of academic geographers (Figure 5.2).

This rapid growth was achieved at the enormous cost of loss of university autonomy. For 'African universities are subject to considerable interference by governments. ... [T]he government usually appoints the Vice-Chancellor.

It also sets the salary scales – often at levels which drive away the most competent – and the budget, with little regard for the cost of teaching the increased numbers of students. It opens new universities for political reasons, regardless of its inability to maintain the old ones. It decides on fees and loan programmes, which are very unpopular with students. Ministers can change the school year and the requirements for university entrance. They usually demand more "relevance", translated as more science and engineering students and fewer in the arts and social sciences, or demand a halt to the production of graduates for whom there are no jobs' (Peil 1995: 101). Much of this is, of course, reminiscent of the consequences of the transformation of the British university sector since the 1980s discussed in the preceding section. Similar grievances from academics in other parts of the world have been reported (e.g. Myers 1995).

Within planning, policy, development and environmental circles, geography's perceived strengths of synthesis and problem solving command premium ratings. Thus, as is often the case in the United Kingdom, the subject remains popular due to a perception among students that it offers better job prospects than competing disciplines (Okafor 1989; Sant 1992). Consequently, as university enrolment has grown rapidly, the number of students offering geography has tended to grow too. Most recently, the subject has been represented in multidisciplinary development and environmental studies units. Some of these units have even, on occasion, been headed by geographers (Sindiga 1988).

While it manifestly cannot be reduced to just this, the current telling of the story of African academic geography privileges the role of 'geographers in high places', whose professional achievements are widely believed to have enhanced the status of the discipline and, by extension, its ability to 'hold its place' within both academic circles and the wider society (Barbour 1982; Peil 1995). Barbour, for instance, considers that the election of the Nigerian Akin Mabogunje, arguably the continent's most 'high-profile' geographer, to the presidency of the International Geographical Union in the early 1980s was both a great personal achievement and an indication that African geography and geographers had clearly come of age (Barbour 1982). The writing of a comprehensive history of academic geography in Africa from an African perspective remains a crucial and increasingly urgent task.

Although poorly represented in elective political office, geographers have long played 'establishment' roles in environment, and regional and rural development policy, planning and management (Howes 1989; Sindiga 1988; Sindiga and Burnett 1988). Maybe as many as a dozen (mostly human) geographers have acceded to vice-chancellorships of Anglophone tropical African universities since the early 1960s. But this particular telling is instructive too for the (admittedly limited) insight it gives into gender relations. Like its Anglo-American counterparts, African academic geography's story is rarely about women in high places (or, indeed, about women at all)! Thus the first women to accede to full professorships in Nigeria and Sierra Leone did so only in the 1980s. To my knowledge, no female Ghanaian geographer has yet been able to obtain a personal chair.

Women remain under-represented in African academic geography, particularly in its senior ranks. This situation has apparently still not been subjected to feminist critiques of the kind that are now commonplace in Anglo-American geography, and that exist within the wider social sciences in Africa (e.g. Imam and Mama 1994). It is, however, symptomatic of a wider problem. For instance, a recent survey of women in public life in Ghana revealed that only one of Ghana's five universities had any women at the professorial level, and that women constituted only 0.5 per cent of the population of professors in the country (ISSER/DPPC 1998). And in Nigeria,

> [t]here is ... discrimination against women in terms of the application of the conditions of [university] service. In matters like housing allocation and annual leave entitlements, women are often treated as appendages of their husbands, where both of them work in the same university.
> (Mustapha 1996: 110)

In stark contrast to the institutionalisation of academic geography in Britain, where it was heavily implicated in both politics and imperialism (Dodds 1994; ÓTuathail 1992), the progress of academic geography in Anglophone Africa owes much to the fact that disciplinary tools, concepts and methodologies are widely perceived as being applicable to everyday problems (Barbour 1982; Howes 1989; Kenzer 1993; Okafor 1989). Thus, in Nigeria as elsewhere, 'distinctively [geographic] voice[s are] characterised by an "emphasis on relevance to national development and ... a trend towards professionalism"' (Sada 1982; cited in Okafor 1989: 212). Indeed, according to the Zimbabwean geographer, Lovemore Zinyama (1988: 59), 'the research endeavours of academics ... should be aimed at improving the level of well-being of their people and society'. The increasing range of 'applied' or 'professional' content of university curricula across much of Anglophone Africa suggests that the practice of geography does reflect the continent's environmental and development problems (Howes 1989).

Such a perspective of geography undoubtedly resonates with the vast majority of African governments, and a large number of Anglo-American and African academics, regarding the role of both geographers and universities within the process of national development (Batterbury 1997; Schroeder 1997; Simon 1997). However, when uncritically promoted, this can become a purely 'instrumentalist view ... which constrain[s] intellectual efforts within the narrow limits dictated by the needs of "development"' (Mustapha 1993: 1). As Zinyama concedes, such a focus has not necessarily encouraged geographers to make widespread use of 'the more radical methodological and philosophical perspectives ... in the examination and interpretation of economy and society' (1988: 68–69). A notable exception here would be the kind of critical geography that emerged in the last decade or so of apartheid in South Africa. Also, current theorisations of economic stabilisation and structural adjustment have stimulated much critical debate within both geography and wider academic circles. More generally, however, could such concentration on the local and national have any bearing on why (cross-country) comparative studies are so rare in Anglophone African human geography? Or why until relatively recently had so few African geographers undertaken 'foreign area' research in Africa? On a personal note, I have myself frequently been chided by compatriots for devoting all my recent research and publishing endeavour to a 'foreign' or 'adopted' (even if African) country. Nonetheless, there is little question that geography's relative status as an academic discipline in Anglophone Africa has been steadily enhanced since the 1940s (*cf.* Dawson and Hebden 1984).

SAPping Times

Since the early 1980s, however, this reinforcement of geography's place within African academic circles has occurred against a background of persistent economic decline. The scale and complexity of this crisis dwarf that currently facing Anglo-American (Africanist) geography (Wilbanks 1993). While university enrolment has increased by almost two-thirds, higher education's share of national budgets has, with a few exceptions, dropped significantly (Kinyanjui 1994). In large measure, the capacity of governments to continue to fund education, even at pre-1980s levels, has declined with deteriorating national economic performance (Land 1994; Neave and van Vught 1994). Surprisingly, however, a generous contact leave scheme (which is separate from normal sabbatical leave) for staff of the University of Zimbabwe has so far survived despite that country's higher education funding crisis and deteriorating exchange rate.

In Nigeria, which accounts for close to half of all Anglophone universities and geography departments, capital allocation to universities dropped from 210.5 million naira in 1980 to 8 million naira in 1987. In the latter year, the real exchange rate value of the local currency against the American dollar was only 29 per cent of its 1980 value of about 1 naira to US$1.80 (Bangura 1994). A depreciating currency, allied with reduced public sector support, spelled disaster for both tertiary education and academics. This was exacerbated by spiralling inflation; irregular, late, and non-payment of salaries; and the absence of effective price controls. The US dollar equivalent of an average senior lecturer's annual salary in a Nigerian university declined 19-fold over the decade to 1993, while the annual rate of inflation increased 24-fold (Bangura 1994; Mustapha 1996). Similarly, while a Ugandan lecturer's monthly salary bought no more than a week's supply of food, a Sierra Leonean counterpart needed to part with a week's wages for the 'luxury' of obtaining an imported paperback (Currey 1986). A Ghanaian colleague wishing to buy a new refrigerator needed to raise the equivalent of a year's salary (Land 1994). Academics were traumatised by this erosion in their living standards and working conditions (Bangura 1994).

As it has become increasingly difficult for universities to compete in national and regional labour markets, the 'deprofessionalisation' (Mustapha 1993) of academics has become widespread. Many academics juggle consultancies and other jobs to make ends meet or have left university employment for more remunerative private sector and international appointments, or government service (Peil 1986). In universities like Makerere and Zimbabwe, up to half of the established positions are unfilled, as African tertiary education loses the fight to keep (some of) its brightest and best brains on its campuses (Land 1994).

The constraining effect of ubiquitous economic restructuring and structural adjustment programmes (SAPs) on social expenditure, in particular the shift in spending away from tertiary to basic education, has repeatedly been implicated in the ongoing transformation. Yet through its assistance for university sector reorganisation, the World Bank has also actively supervised the restructuring of African tertiary education and, through this, directly influenced the transformation of the conditions under which academic geography is produced. Throughout the 1980s and 1990s, 'education lending was linked much more explicitly to policies for macro-economic adjustment' (King 1990: 52). In the absence of alternative visions of educational development considered acceptable by the bank, either from donors other than itself or from African sources (see, for example, Bako 1994), the bank's neoliberal agenda has prevailed (King 1990). Indeed Donald Ekong, secretary-general of the Association of African Universities, was forced to observe that even though no higher education institution has gone into receivership, a lot seem to be run by international agencies (Iloegbunam 1993).

This peculiar expression of global–national–local articulation has perhaps proceeded furthest in places like Ghana and Nigeria. In the latter, as elsewhere, the objective appears to be to rebuild established institutions in the context of new 'market-driven structures of incentives' (Bangura 1994: 299). Although precise details vary within and between countries and universities, overall, concessionary loans have provided capital for equipping libraries and laboratories, funding research, etc., in return for programme 'rationalisation'. This has involved staff retrenchment, disciplinary and departmental mergers and closures, increased cost sharing, and, in some cases, a shift of emphasis from arts subjects to science- and technology-based disciplines and education (Adepoju 1993; Neave and van Vught 1994).[6] Yet despite the introduction of greater cost-sharing arrangements, there has not been any significant improvement in working conditions in universities (Gana 1993; Iloegbunam 1993; King 1990; Logan and Mengisteab 1993; Mustapha 1993; Samoff 1990). Well into the 1990s, for instance, cutbacks were only partially made up by replacement funds earned by widely and hastily established university consultancy units and other commercial ventures, which had emerged as part of sectoral reform (Land 1994; Mustapha 1993).

Teaching loads and administrative responsibility remain unacceptably

---

[6.] For specific ways in which the bank and other donors directly affect education policy and practice, see Bako (1994) and McGinn (1994). The volume edited by Neave and van Vught contains a detailed study of these processes at work in Ghana (Neave and van Vught 1994). Bangura (1994) discusses the Nigerian case. Also, it is worth recalling that (1) internal university mechanisms were effective in 'resisting aspects of [reform] programmes and refining others in such a way as to preserve what the universities consider the minimum conditions for the maintenance of their integrity as institutions of higher learning'; and (2) the introduction of reform 'and the vigour of its prosecution have driven the universities to take some positive measures, many of which have been under consideration internally for years' (Sawyerr 1994: 47–48). Although they refer specifically to the Ghanaian case, these observations must surely be of wider applicability.

heavy, classroom and laboratory equipment woefully inadequate, and library resources rudimentary or dated. As in 'traditionally black colleges in America ... limited budgets result in heavy teaching loads and little time for scholarly research' (Cohen 1990: 232). In addition, disruptions to normal university activity by student and staff strikes, as well as civil disturbances, which were common throughout the 1980s, 'made it impossible to plan any systematic research agenda' and discouraged 'sustained speculative thinking' (Mustapha 1993: 15). Recent events in Kenya, Nigeria and Niger, among other places, suggest that these observations remain valid in the late 1990s. Many universities continue to economise on journal subscriptions, cut back on book imports, restrict local and overseas conference attendance, and drastically curtail funds available for local research. In a leading Nigerian university, staff and student access to the library is restricted to a single day a week in a bid to cope with overcrowding (Mittelman 1997).

Differentiating transformation impulses

This gradual erosion of both the institutional structures and spirit of tertiary education has occurred in complex, uneven and frequently contradictory ways. Thus university administrators, who have frequently been locked in conflict with academics on their staff, are themselves demanding an end to state monopoly of higher education, and greater managerial and administrative autonomy, 'to enable them to generate resources for self-development' (Land 1994). They also want a review of long-established policies that allow undue political interference in educational institutions (Anon. 1993). These measures are also widely supported, and in many cases were initially advocated, by academic staff unions across the continent. For an interesting representation of how university autonomy could, on occasion, be almost sacrosanct in the 'good old days', see Sawyerr (1994). At the same time, in a manner reminiscent of David Smith's characterisation of the competition for funding in British geography, inter-university competition for available consultancy contracts has been sufficiently fierce to threaten collegiality on occasion (Smith 1994). He speaks of such competition mirroring the business world and its ethics, and of departmental heads acting more like entrepreneurs than scholars. On a related note, Imam and Mama (1994: 76) speak of 'despicable' aspects of the 'consultancy syndrome' among African academics, which include a 'willingness to do anything for money, to make recommendations in areas in which one has no expertise rather than suggest someone else, to do bad and shoddy and what one knows to be useless and reactionary work, for the sake of benefits to oneself alone'. I have heard academic colleagues in at least two of the larger Anglophone countries debate the (de)merits of zoning national space into 'consultancy catchment areas' as a mechanism for regulating such competition in their individual countries.

Similarly, within individual institutions and departments, friction has frequently developed over the 'politics of exclusion', which reportedly influences the selection of researchers to work on remunerative consultancy contracts (Cline-Cole 1996). In some cases, these conflicts pit junior academics against their senior colleagues, whom they accuse of monopolising the limited opportunities available for professional and personal advancement (Porter 1995). In others, they fuel interdisciplinary and interdepartmental rivalry (Cline-Cole 1996). Increasingly, too, there is wider acknowledgement of the fact that transformation impulses

> are disproportionately felt by women academics, particularly in that they are wives and/or mothers. [And t]he additional burdens being borne by women are not often raised in discussions of academic freedom or social responsibility of intellectuals.
>
> (Imam and Mama 1994: 76)

Significantly, the consequences of transformation have been no less uneven with specific reference to geography and geographers. The discipline does not appear to have been one of the most adversely affected by restructuring, which has reportedly been used by some governments to victimise non-compliant disciplines like political science, sociology and history (Gana 1993; Mustapha 1993). It was quite clear that successive military governments in Nigeria, for example, used the opportunity of university sector restructuring to target disciplines and lecturers who, in official-speak, 'taught what they were not paid to teach'. This raises important questions of academic freedom which are beyond the scope of this chapter. For useful treatments of academic freedom, see Diouf and Mamdani (1994), CODESRIA (1996) and Mittelman (1997).

It is not entirely surprising that geography was targeted less than some other disciplines. Not only was the mainstream focus in African geography seemingly apolitical, but geographers were also not generally prominent as either political activists or government opponents. Indeed, well after the crisis had set in, new geography, geography-based (e.g. land resources at Bayero University, Kano, Nigeria) and geography-related (e.g. environmental studies at Moi University, Eldoret, Kenya) courses and programmes were still being established or strengthened, sometimes with financial assistance from donor agencies like the Ford Foundation and the European Economic Community (now Union) (Mortimore *et al.* 1987; Paul 1993; Sindiga 1988). Undoubtedly, the flexibility offered by geography's multiple-faculty affiliation contributed to such outcomes by making it possible for subdisciplines and specialisations in demand to expand and, whenever possible, to exploit interest, goodwill and funds. Equally certainly, such outcomes were also facilitated by a combination of geography's previously identified 'development bias', and the influence of 'geographers in high places'. The latter were instrumental in channelling research and consultancy contracts in the direction of erstwhile academic colleagues/friends.

In contrast, the actual practice of university geography fared less well. Laboratory teaching and research have suffered inordinately, as have field courses. In particular, survey and cartographic equipment and material have frequently been in acutely short supply. In some of the worst-affected departments and universities, it has been impossible for highly trained and accomplished cartographers to produce camera-ready diagrams because of a lack of building and equipment maintenance (*cf.* Barker 1986). In many cases, the practice of geography has become what geographers *can do* with available resources, rather than what their professional training and judgement tell them they *ought to do* and would dearly *like to do.*

More generally, there has been a marked decline in local and regional professional contact and communication. At the start of the 1970s, Anglophone West Africa boasted three geography journals, in addition to at least two other regional journals favoured by geographers. By the mid to late 1980s, however, only one of the geography journals (and two of the original five) was still being published. Similarly, although the Nigerian Geographical Association (the largest in tropical Africa) continues, to its credit, to organise regular annual conferences, reduced attendance and limited funds have forced it to replace the 'traditional' publication of edited proceedings with the more 'pragmatic' response of a simple compilation of conference abstracts. Elsewhere, and albeit for different reasons, even South African geography, which, throughout the apartheid years, succeeded in maintaining strong academic links with Anglo-American geography, has been remiss in not extending its post-apartheid gaze sufficiently quickly to encompass geography and geographers north of the Limpopo River (Simon 1994), although this process is now underway (David Simon, personal communication). As mentioned earlier, however, the University of Zimbabwe has managed to retain its staff contact leave scheme to date.

Overall, geographers and their discipline continue to create and exploit spaces for manoeuvre wherever and whenever opportunities present themselves. It is a capacity that is proving indispensable in the ongoing struggle to respond adequately to threats to sustainable livelihoods posed by economic stabilisation and structural adjustment, and the restructuring of African tertiary education. Geographers and other academics who remain in university employment have responded to salary differentials and varied career prospects in much the same way as their Anglo-American counterparts have always done. For instance, the much greater degree of mobility of practitioners that has characterised British geography during the 1980s and 1990s (Sidaway 1997a) was observable in West Africa much earlier. Here the 1970s' petroleum boom fuelled a massive expansion in tertiary education in Nigeria. As local salaries were competitive, even by European and North American standards, this encouraged a 'brain drain' from other West African countries and much further afield (Swindell 1990). By the mid- to late 1980s, however, migratory flows reflected the dramatically altered economic fortunes of West Africa's largest economy. Academics (including Nigerians) were moving to East and southern Africa, the Arabian Peninsula, and the North. Even a cursory examination of the *Commonwealth Universities Yearbook* and *Orbis Geograficus* reveals how well represented geographers were in all these flows. Ghanaian and Sierra Leonean geographers, for example, continue to be listed in universities in countries as far apart as Nigeria, Kenya, Botswana, South Africa, Britain and the USA. Since the 1980s, moreover, many Third World geography students who obtained postgraduate degrees in the United States did not return home at the end of their studies (Paul 1993). Within Nigeria itself, academics have always moved (and continue to do so), whenever this is perceived as a good career initiative, from older to newer universities and from universities in various parts of the country to those located in or near(er) their states of origin (Peil 1986).

Further, along with other public sector employees on fixed salaries whose real earnings have declined since the 1980s, university lecturers have had to exploit a diversity of informal sector opportunities 'as a means of containing, possibly reversing the obvious slide in their living standards' (Mustapha 1993: 21). In Tanzania, they 'invest increasing amounts of time and effort in consultancy work, and in informal sector activities like rearing chickens, running pig farms and dairy cattle units, driving taxis and pick-ups or supervising their [mini] buses' (Lugalla 1993: 205). Farming, retail and petty trading and small-scale manufacturing represent important additions to this list of income-earning activities in Nigeria (Gefu 1992). Indeed, 'the [Nigerian] government is allegedly worried about the extent of the involvement of university teachers in private businesses, and the National Universities Commission (NUC) has accused such teachers of "double dealing"'(Mustapha 1993: 21). The irony inherent in this is, I am sure, not lost on academics who, after decades of being associated in the popular imagination with self-isolation in so-called 'ivory towers', now find themselves criticized by government for being worldly-wise. Hagan (1994) provides a wonderfully entertaining caricature of the absent-minded professor in Ghana.

Ironically, therefore, as academics have responded to the subsistence crises they face, and, in the process, neglected university responsibilities, they have left themselves open to charges of a lack of professionalism. In reality, however, it is less these livelihood activities *per se* and more the underlying socio-economic and political conditions to which they respond that represent the direct enemy of professionalism (Porter 1995). Their persistence into the late 1990s makes the pursuit of any intellectual endeavour or semblance of academic 'normalcy' in the African tertiary education sector all the more remarkable and commendable (Hargreaves 1985; Wilbanks 1993). This is doubly so, as many academics, geographers and non-geographers alike,

> have simply given up any hope of any systemic solutions. They instead seek solace in client networks in the university, and particularistic ethnic and religious groups which offer a modicum of solidarity and support. Many others have found it profitable to collaborate, individually or through their professional associations, with the powers-that-be in the society for some pecuniary advantage.
>
> (Mustapha 1993: 14)

Nonetheless, the most serious consequence of prevailing socio-economic conditions for African higher education may well not be the overall decay of intellectual and professional life *per se*. It is arguable that academe runs a real risk of becoming marginalised in the production of research knowledge and data that impact seriously on the 'development' of the societies of which it is an integral part (Kinyanjui 1994). Indeed, not only are African intellectual voices too rarely heard, even within Africa itself, but also it is precisely within the resulting intellectual and institutional void that neoliberalism has become the hegemonic 'development' discourse (Hyden 1996).

In African as in Anglo-American geography, then, practitioners are faced with the common task of consolidating their discipline's changing status, both within academic circles and the wider society. Yet the two sets of practitioners are having to contend with profoundly different structures of opportunities and constraints. Much of the logic of 'new' interaction thinking and practice appears to be organised around questions of how and why professional exchange ought to be deployed to maximum mutual benefit under these circumstances.

Locating ('new') interaction in a transforming Anglo-(American) African(ist) geography

Although the historical background to contemporary academic exchange lies in the colonial origins of tertiary education in Africa, and geography's wider implication in colonialism, it is the intersection of global political economy and Anglo-American and African tertiary education that provides the immediate context (Driver 1992; Hargreaves 1985; Johnson 1989; King 1990; Morgan 1992; Peil 1986; Samoff 1990; Smith 1994). Furthermore, this context needs to be located on the broader canvas of the growing polarity of world income, which has failed to enhance Africa's overall position in the international division of labour (Cliffe and Seddon 1991; Samoff 1991; Adedeji 1994; Peet and Watts 1993; Himmelstrand *et al.* 1994; Adepoju 1993). However, the production of academic geography is simultaneously a process of intellectual and academic reflection and a labour process. Thus it highlights contradictions, even as it underscores the ways in which interests coincide/coalesce across localities and institutional spaces.

Interaction in practice

Figure 5.3 shows that the structures and practices of North–South interaction bear the clear imprint of gross unevenness in global patterns of academic (and other) 'development'. But it also hides the bewildering array of rationales, motives, mechanisms and modalities which characterise such professional interaction. For instance, a much larger proportion of the exchange portrayed is undoubtedly mediated through brokers' networks of solidarity/trust than within the context of large formal linkage schemes or collaborative programmes, which combine several of the flows depicted in separate but interlinked components (Wilbanks 1993; Smith 1994; Howes 1989; Sindiga and Burnett 1988). In any case, some of the exchanges (including many British Council-sponsored schemes which

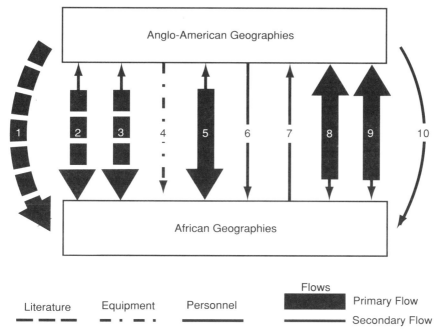

Literature Equipment Personnel

Flows

Primary Flow

Secondary Flow

1 Published books 2. Journals 3. 'Grey' literature/personal correspondence
4. Teaching/laboratory/office supplies and equipment 5. Expatriate geographers
(teachers) 6. Field research and classes (teachers and [research] students)
7. Advanced studies by African students 8. Exchange of geographers on sabbatical,
fellowship etc. 9. Conference participation/study tours/professional
meetings 10. External examiners

Figure 5.3 The contours of professional interaction in Anglo-(American)
African(ist) geographies in the 1980s and 1990s
Source: Adapted from Paul (1993: 463)

link British geography departments with their African counterparts) are
'less a vehicle for significant project aid than for maintaining academic in-
terchange in situations where access to routine sources of materials,
journals and equipment is problematic' (King 1990: 55). Much 'new'
interaction merely advocates an intensification of relations of this kind
(Paul 1993; Smith 1994; Wilbanks 1993). Academic geography is also
produced within the context of affiliations/interaction involving
development aid projects (Batterbury 1997). I return to the subject of
networks of solidarity/trust and their role in North–South interaction
below.

Anglo-American resources already underwrite many of the flows depicted in Figure 5.3, although a small number of African universities
maintain fairly adequate travel and library purchase funds, which allow
them to contribute to the costs involved. For example, journal distribution
programmes and book collectives provide subscriptions to journals and
arrange the collection and transportation of donated books to university

and research libraries in Africa (flows 1 and 2). Travel grants support the attendance of African academics at international meetings (flow 9). The exchange of geographers (flows 5, 6, 8 and, to a lesser extent, 7) is funded overwhelmingly by Northern resources. South–North flows of literature and personnel (particularly graduate students) were much greater in the period prior to the 1980s and 1990s. Not surprisingly, in the years since the 1980s, and seen from the perspective of African geographers and geography departments,

> never before have [Anglo-American] geography departments been so well equipped, apparently (that is to an outsider) bulging at the seams with micro-processors, word processors and laboratory and field equipment. Never before has the resource gap between geography departments in the industrial nations and those in the Third World countries been so great.
>
> (Barker 1986: 157)

But there are enormous variations in the extent and intensity of participation in interaction. Thus the Association of American Geographers' journal distribution initiatives for sub-Saharan African universities and departments of geography is not matched by similar initiatives on the part of the Institute of British Geographers (IBG) – now merged with the Royal Geographical Society (Paul 1993; Smith 1994; Wilbanks 1993). However, the Developing Areas Research Group (DARG) of the institute recently initiated both a Postgraduate Research Award and an Undergraduate Dissertation Prize to encourage interest in the study of developing areas by students of all nationalities registered in British institutions of higher education. DARG also plans to distribute free copies of its monographs and new series of commercially published textbooks on development in different regions to universities in the poorer developing countries in exchange for material published locally in these countries. Similarly, while Nigerian universities have long only used Anglo-American geographers as external examiners in exceptional circumstances, the University of Sierra Leone still does so on a more regular basis. More recently, as part of a drive to establish/maintain 'international' post-apartheid standards, South African universities appear to be institutionalising the practice of using 'overseas' geographers (generally from North America, Europe or Australasia) on postgraduate examination panels (David Simon, personal communication).

With reference to African academics, those who remain in the field try to exploit their professional research skills and networking abilities. Some have (re)oriented their research to service the needs of government and the private sector, while others 'tap international opportunities, and become consultants to multilateral and non-governmental agencies or establish linkages with international research networks' (Bangura 1994: 274). Athough all research funding agencies have their agendas, research within the context of international exchange networks such as the Joint American Council of Learned Societies–Social Science Research Council Programme on African Agriculture (ACLS–SSRC) and the Global

Environmental Change Programme of the British Economic and Social Research Council (GEC–ESRC) often provides greater autonomy in the selection and treatment of research topics, than research 'commissioned' by NGOs, donor agencies and governments (SSRC 1989; Wilbanks 1989). This mode of professional interaction represents one of the few remaining options for academics wishing to undertake adequately funded critical research. Furthermore, as with the ACLS–SSRC programme, the maximisation of African participation is frequently a declared programme goal.[7]

But if the success rate among first-time Anglo-American grant seekers for research funding of this kind can be as low as 20 per cent, even lower rates have been recorded among African social scientists (Dando 1989; SSRC 1989). Intellectual isolation limits knowledge of available grants and current grant-winning ideas among experienced African geographers. At the same time, inexperienced young(er) (and particularly) locally trained African geographers rarely have adequate understanding/ knowledge of international funding sources and announcements, skillful proposal writing and grant administration experience (*cf.* Dando 1989; SSRC 1989). This seriously constrains their ability to write compelling proposals rooted in prevailing (Anglo-American) theories. Of particular relevance here is that these theories appear to have increasingly short 'shelf lives' (Mohan 1994). But even this still leaves unresolved questions of who decides (and on what or whose authority) which issues are (not) compelling, and why, how and to what end (see, *inter alia*, Kay 1993; Mohan 1994; Sidaway 1992; Smith 1994). Yet, there appears to be little prospect for immediate change. Guidelines for successful grant-seeking continue to specify requirements which African geographers operating with grossly inadequate resources can fulfil only with extreme difficulty, even assuming they had prompt access to these guides in the first place (see, *inter alia*, Abler 1988; Abler and Baerwald 1989; Dando 1989; Wilbanks 1989). A critical 'new' interaction could provide some respite by

[7.] In the United States, while the applied research programmes of mission agencies and the National Science Foundation are the largest potential sources of funds for geographic research (Wilbanks 1989), more Africanist geography research is probably *actually* funded by private and/or charitable organisations. In Britain, the Economic and Social Research Council (ESRC) and Natural Environment Research Council (NERC) still provide most of the research and training funds that allow academic geographers to continue to work overseas, even if the proportion of their overall budgets devoted to either overseas or geographical research and studentships remains low (Lawler 1992; Potter 1993). The ESRC's 'African connection' – through its graduate studentships and GEC programme – is the better developed of the two, with a quarter of the fifteen projects funded under GEC Phase 1, for instance, being devoted to tropical Africa (ESRC, nd). Similarly, through programmes like its Life Sciences and Technologies for Developing Countries Programme (STD3), the European Commission is becoming increasingly influential in encouraging and funding collaborative research between European and African universities (EEC 1991). For the Third World as a whole, however, the major sources of research funds in recent years have been the World Bank and the IMF, and the US Agency for International Development (Paul 1993). It is worth mentioning the understated role, even within Africa itself, of the Council for the Development of Social Science Research in Africa (CODESRIA, based in Dakar) in facilitating research, promoting research-based publishing, and the creation of forums for the exchange of views/information among African researchers. African geographers have been slow in recognising (and making use of) the opportunities offered by CODESRIA.

creating space for African academic voices to be heard more insistently and more widely (Hyden 1996; Smith 1994).[8]

Ties of 'reciprocity' that truly bind?

In a whole range of complex ways, then, the continued survival and growth of both Anglo-American and African geography are intertwined with the fortunes of 'distant others', and the conditions under which they produce geography. Within British geography, recent transformations are currently 'normalising' the frantic struggle to 'publish or perish', which has long characterised the professional existence of (especially young) academic geographers in America (Haigh 1982). Indeed, in terms reminiscent of Haigh's references to a historically 'volatile' and 'very much alive' American geography, Sidaway's (1997a) interviewees speak variously of how competitive research funding and periodic research assessment exercises have forced academics 'to get their act together', created a more overtly 'cut-throat' work environment, and stimulated an 'amazingly vibrant' discipline. Successful 'new' institutional interaction can contribute enormous benefits to such dynamism, particularly in the area of research (Porter 1995). But this may be driven predominantly by departmental/university 'pecuniary self-interest' (Smith 1994: 365). For from many a Northern institutional perspective,

> intra-European links as well as recent link developments with Eastern Europe and the Soviet Union promise a greater degree of academic symmetry than some of the links with developing country institutions. In these harsher times when research productivity is being assessed much more seriously in countries like the U.K. ... Northern institutions are likely to prefer relations with the South which are more nearly symmetrical, or where this is not the case, which allow significant research and consultancy to be carried out by the North. What will be hardest to justify will be asymmetrical links involving tutorial and training relations, unless the latter bring to the North major income in terms of overseas student fees.
>
> (King 1990: 55)

Clearly, the marginalisation of Africa impinges upon, and threatens, African(ist) scholarship (Gana 1993; Hyden 1996; Kinyanjui 1994; Mustapha 1993). Thus the ability of Africanist geographers to compete effectively in the new Anglo-American disciplinary/institutional 'marketplace' turns in crucial ways on the maintenance and even expansion of the North–South professional contacts in Figure 5.3. For these geographers, professional/

[8.] It is significant therefore that the JACLS–SSRC addressed this problem directly. It organised workshops (led whenever possible by African academics employed in American universities) to teach young African academics the skills of proposal writing, and screened proposals from collaborative teams of African and Western scholars to 'ensur[e] that all team members have equal intellectual leadership and carry equal responsibility for fieldwork and analysis' (SSRC 1989: 6; Rick Schroeder, personal communication). Also, each of the fifty-eight projects funded between 1987 and 1989 had an African institutional base. For a related point about the value to Third World academics of involving them fully in all stages of the design and elaboration of collaborative North–South activity, see Porter (1995).

personal advancement is as heavily dependent on the creation and nurturing of these structures and relationships, which ensure (in)direct access *to* 'the field', as it is on actually being *in* 'the field'. Consequently, individual geographers have long invested generously in the social capital represented by Africa-based or -focused networks and reputations of trust that bridge state, market and civic spheres, on the one hand, and inter- and intra-university spaces, on the other (*cf.* Bebbington 1997). Not surprisingly, people like David Smith urge resistance against those elements of the current transformation that would seek to discourage/minimise the moral, intellectual and practical value of investments of this kind:

> If ... pecuniary self-interest is now our predominant motivation, this will discourage us from doing things which are unlikely to appeal to those responsible for research rating. For example, we will avoid compiling edited collections of papers, even if these facilitate the publication of work of people hitherto constrained in the dissemination of their research. ... We will avoid the time-consuming editorial work required to publish a Russian colleague's book in English ... for which we may earn no UFC credit. ... We will not publish in most foreign journals, which the research raters may deem inferior to our own. And it would be unwise spending time organising international seminars, helping visitors from overseas and attending less prestigious conferences elsewhere, unless we can see some tangible 'product' to put on the next UFC forms. We may even be discouraged from writing volumes on distant people and places.
>
> (1994: 365)

In any case, the major funding agencies that support research in the South do not just 'require Western consultants for their expert advice [they also] encourage collaboration between researchers of both developed and developing countries' (Paul 1993: 462). Indeed, many proposals for foreign area research are likely to be turned down, their intellectual merits notwithstanding, if they make no allowance for host-country partners (Wilbanks 1993). The point needs to be made that this now taken-for-granted insistence on Southern participation in Northern-funded projects is itself an outcome of recent preoccupation with professional ethics within academic circles and the development sector (Proctor 1998). This has major implications in an era in which 'finding, obtaining, and managing money from external funding sources to do one's research, is one of the new job skills needed by professional geographers' (Dando 1989: 3). It suggests that the future of Northern-based foreign area research may well lie in greater collaboration 'in the field' with Southern academic colleagues, who have valuable intellectual and practical insights to contribute to such a shared project. From both a practical and a moral point of view then, Anglo-American geographers

> need to know geographers in other parts of the world, because without them we will know too little about those areas, because without them we may be unable to work there ourselves.
>
> (Wilbanks 1993: 2)

'New' interaction emphasises the perceived mutually beneficial nature of such interdependence. Its main theme is that greater interaction, *can* bring material and intellectual resources to both Northern and Southern geographies. Thus in the case of the University of Khartoum, not only have 'the [geography] faculty's contacts with government officials and nonprofit organisations ... provided [Anglo-]American researchers with invaluable contacts for research activity', but formal links with British and American universities 'have resulted in Sudanese students obtaining postgraduate degrees from those institutions, in addition to valuable equipment acquisitions for the UK department' (Howes 1989: 217).

In intellectual terms, a 'new' interaction would introduce many more Western geographers to 'valid perspectives and world views of geography [other] than just those current in Europe and North America' (Haggett 1990: xvii, cited in Paul 1993), while at the same time providing Southern geographers with access to 'recent technical, conceptual, and philosophical advances in Western geography' (Paul 1993: 46). This should contribute to countering ethnocentrism. Viewed in more prosaic terms, however, 'new' interaction is about extending the social capital that is already inherent in brokers' networks of solidarity and trust, to our professional organisations/associations and our institutional homes (*cf.* Bebbington, 1997). To this end, it reserves a privileged role for individuals/groups/networks with specialist interest in North–South exchange and/or knowledge of the context of the production of geography in both areas. These brokers include regional speciality groups as 'windows to their respective parts of the world' (Wilbanks 1993: 2); researchers from the Third World now employed in First World institutions; geographers from developed countries with research interests in developing countries (Porter 1995: 141; Paul 1993: 464); and Third World national geography associations/commissions. Significantly, too, although no target individuals are identified within African geography, a mix of mostly young and mid-career academics appears to be particularly well represented in South–North personnel flows in Figure 5.3. Without these networks, and the favourable institutional environments that they seek to create/maintain, the task of establishing 'meaningful' professional dialogue will be made considerably harder (Bebbington 1997; Porter 1995; Smith 1994; Wilbanks 1993).

Nonetheless, networks of this kind are simultaneously inclusive and exclusive, empowering and disempowering. Okafor's (1989) review of research in Nigerian human geography, for example, omits most of the research from the northernmost parts of the country, including scholarship recognised as influential within the wider social sciences in the country. This is, of course, reminiscent of Berg and Kearns' (1998: 129) observation of the role of banal disciplinary practices in the discursive constitution of geography, and of how 'unlimited and unmarked geographies ... mark out, constitute, and limit the geographies of the Other'. A much clearer idea, and more detailed understanding of their sociology and geography ought, therefore, to be central to any 'new' interaction (Cline-Cole 1996).

Constructing 'interactive' meanings

Asymmetry plays a central role in structuring 'relations of reciprocity' among academic geographers. In particular, it has implications for negotiations over power/knowledge in African(ist) geography, and in African Studies more generally. Why, for instance, have so few African geographers ever served as external examiners and programme assessors in Anglo-American Africanist geography? Why has the long history of field research in Africa by Anglo-American geographers not been matched to anything like the same extent by the reciprocal study of Anglo-American landscapes and societies by African geographers? Why do so few articles by African geographers in Africa ever make it into the pages of the major Anglo-American journals? Why do even fewer quality articles by Anglo-American geographers appear in African journals? Why are publications in Anglo-American outlets still accorded greater prestige than local publications in many African universities? Why are the African authors I have been able to draw on most extensively for this paper those who, like me, are expatriated from both their home countries and continent? Why does it seem so much easier to write about Africa as a whole from any British or American university town than from, say, Kano, Freetown or Eldoret?

The answers are no doubt enormously complex and implicate a whole range of individual/departmental/institutional circumstances, which have to do with both agency and structure. For instance, the exigencies of day-to-day living leave home-based African colleagues little time or resources for scholarly pursuits, particularly for academic publishing, which is itself subject to its own internal economic/political logic (Hyden 1996; Porter 1995). On the other hand, as Simon explains, expatriation can provide security and the opportunity 'to engage in reflective debates' without these stresses of daily living, frequently in an intellectual environment enriched by influences and 'fellow migrants' from various parts of the world (Simon 1997: 236, note 13; Chapter 2, this volume). Nonetheless, there is a widespread perception among African academics that the sheer scale of this disparity possesses real potential for silencing African voices via the appropriation of the means of knowledge production and dissemination in African Studies by Anglo-American academics (Mustapha 1993). Indeed, such a perception apparently led to a call by the 1985 Conference of African Vice-Chancellors, Presidents and Rectors of Institutions of Higher Education for the 'delinking' of their institutions from American and European universities (Crowder 1987).

In the event, neoliberal impulses are currently reintegrating the African academy into global academe, albeit in newer, even more dependent and more marginal ways (Kinyanjui 1994). Thus, not only is increasingly less of the production of knowledge about Africa occurring within the continent itself, but also the bulk of research on Africa is still being produced in the North, with Anglo-American writing 'reach[ing] furthest and in that sense help[ing to] determine intellectual trends' (Hyden 1996: 4–5). Clearly, however, the danger of a possible turn toward academic autarchy in either the North or the South has not been lost on the proponents of a 'new' interaction in geography or on Africanists who share similar preoccupations. Indeed, one of the latter had cautioned against

> the creation of two separate and compartmentalised worlds of Africanists: [poor] African and [rich] non-African ... with 'us', the comparatively well-off non-Africans, increasingly dominating the field, and 'them', the African scholars, feeling that they have less and less control over the means of production in African studies.
>
> (Crowder 1987: 109)

It is crucial, therefore, for 'new' interaction not to be seen to be reinforcing existing relations of domination and oppression, particularly as professsional interaction, 'new' or otherwise, 'take[s] on different meanings for different actors in particular historical contexts' (Moore 1993: 397). This is hardly surprising. '[T]he search for "meaning" is not conducted in a purely personal, introspective way, nor in a social vacuum, but under definite material and social conditions' (Peet and Watts 1993: 248). Co-operation, conflict and controversy thus constitute the 'realities' and 'facts' of such interaction. These contestations/negotiations remain central to personal/group/institutional experiences of resource sharing. In particular, they help to counter any tendency towards 'the abrogation of genuine geographical knowledge about the ... complex, diverse and heterogenous social mosaic of places [people and relations]' implicated in interaction (ÓTuathail and Agnew 1992: 202).

For instance, a number of academics (including geographers) returning from visitorships to Africa under the American Fulbright Program report a home work environment that either did not value their African experience or was jealous of it. Indeed, some of their colleagues were convinced 'that it had been an easy year off and now it was time to really get down to work' (Sunal and Sunal 1991: 53). Similarly, within the context of resource shortages described earlier, African geographers (and their departments) with well-developed formal and informal links to Anglo-American geography are commonly the envy of many of their peers. They are widely perceived as nascent professional 'aristocracies' with ready access to overseas travel funds, research grants and current literature. Certainly, deans/heads of such departments can translate success at attracting overseas resources into (sometimes undue) influence in central university decision making, presumably to the detriment of other – less 'successful' – administrators, departments and disciplines.

Local critics have been most sensitive to the perceived privileged status of this aristocracy in circumstances where university autonomy is under threat from national government; where state-imposed vice-chancellors lack natural constituencies within the universities they preside over; and where prolonged strikes by university lecturers have resulted in the suspension of salaries, staff dismissals and college closures. In particular, where individual aristocrats are known critics of university and national policy, politicians and university administrators (often one and the same thing) argue that the 'foreign connection' puts such people and their departments beyond effective administrative control and censure, thereby transforming them into (potential) security risks. In some countries, therefore, restrictions long imposed on foreign researchers have gradually been extended to local researchers with strong overseas connections. Thus proposals for extensions to existing departmental/faculty linkages that require official approval become subject to close (often frustratingly long) official scrutiny. Unfortunately, too, peer accusations of unsavoury overseas links (e.g. CIA connections) still possess (admittedly currently much-reduced) potential for destroying professional careers in some African universities.

Success at grant seeking and institutional networking can also have unexpected negative side effects. Within a context of declining university resources, aristocrat departments which are perceived as being 'rich' sometimes find that their share of recurrent university funding is reduced in favour of departments with needs that are perceived by university managers to be both greater and more urgent. In extreme cases, this can lead to the failure or inability of departments to maintain properly equipment and vehicles acquired as part of interaction arrangements. More commonly, it leads to the use of project/programme resources for routine departmental or university business, sometimes to the detriment of the original purpose for which they were obtained.

But reservations about aristocrat geographers and their professional links are not restricted to non-aristocrats. Few aristocrats would deny that their participation in interaction is characterised by extreme ambivalence. Beyond the purely personal rewards of improved career prospects or actual advancement, many are unsure of the true nature of the wider role that both they and their subject are being called upon to assume within the processes of disciplinary transformation and institutional restructuring. Thus some of these geographers see a 'new' interaction as a way for more prosperous Anglo-American geographies to fulfil a moral obligation to less fortunate geographies by assuming a share of the burden of rejuvenating these branches of the wider discipline. Yet others point to the 'tied' nature of some forms of interaction as evidence of 'dynamic, subtle but on-going Western imperialism' (Madge 1993: 297). For instance, Anglo-American support to African university libraries frequently restricts purchases to titles published in the home country of the funding institution/agency. Furthermore, United States Information Service libraries in Africa keep holdings that are exclusively North American. British Council libraries similarly restrict their holdings to titles published in Britain. In such overtly self-serving forms, interaction becomes inseparable from wider aid and economic relations between North and South (King 1990). This generates mistrust of the motives of proponents of increased interaction, and serves to limit the latter's appeal among African academics (*cf.* Kidd 1993).

Anecdotal evidence also appears to suggest that there is a real danger that where the primary role of asymmetrical interaction is that of facilitating otherwise problematic Southern access to materials, journals and equipment, the opportunity for undertaking field research is considered by some 'donor' geographers to be the only material benefit they derive from such 'exchange'. Freedom to carry out field research is thus conceived as an inalienable 'right' that nothing – including requirements for obtaining formal research permission or contributing to teaching in host institutions – is allowed to interfere with (*cf.* Crittenden 1988; Madge 1993; Porteous 1988). In the event, host departments point out that assistance with teaching is of greater institutional value, while visiting researchers counter with the argument that research and publication respond more directly to the professional demands of quality control and career advancement in an increasingly demanding home academic sector. Frequently, the research in question is that which is designed to respond to funding preferences, and to make a favourable impression on research assessors at home. This is not necesarily research that is of much interest to host colleagues or of greatest value to their country's immediate needs (Mustapha 1993: 16; see also Porter 1995; Smith 1994; Wilbanks 1993).

However, research is not contested only at the level of its focus or agenda. Associations with colonial rule and contemporary academic imperialism ensure that perceived 'data mining' by Anglo-American geographers generates feelings of intense animosity among African academics. Indeed, the imposition/persistence of onerous 'research access rules' in a number of countries has been justified on grounds of a perceived need to discourage this practice, 'which enable[s] people from the First World to ... write about people from the Third World in a (prevalently) one way flow of information' (Madge 1993: 295). But there is also its converse, 'piggy-back riding'. Thus a number of Anglo-American colleagues speak of African careers advancing 'on the back' of joint publications, which Africa-based colleagues see for the first time only after these have appeared in print. Such 'con'-joint authorship is problematic, even if born of expediency (the difficulties/pressures of daily existence hinder African intellectual input into collaborative ventures of this kind; *cf.* above). In particular, it gives the misleading impression of creating space for African voices in published debates about Africa, even while it is itself an integral part of the complex processes by which such voices are represented and thereby selectively silenced.

Clearly, here as in the struggles over the place of overseas geographies within Anglo-American geography, professional interaction is an integral part of the social worlds of geographers. These are both uneven and contradictory, and their negotiation demands much critical self-awareness on the part of geography practitioners (Madge 1993; Porter 1995; Sidaway 1992; Smith 1994). A 'new' interaction must stimulate further debate of these contestations, if only because it is itself deeply implicated in the worlds which produce them, and whose geographies they help to (re)constitute.

Conclusion: 'new' interaction dilemmas

['New' interaction] could be a way of reforming, with distant others, the relationships which used to bind at least some of us in mutual collaborative endeavours in which questions of personal or departmental credit seemed inconsequential. It may provide a way of beginning to build a broader collaborative structure, towards a universal professional geographical ethic of care, to challenge and hopefully subvert those forces of darkness turning the practice of geography into an even more extreme expression of hierarchical domination and uneven development.

(Smith 1994: 366)

Lowe and Short (1990) identify three types of geographical progress. *Progress of geographers* is individual achievement that enhances career prospects. *Progress of geography* refers to the discipline's ability to maintain or improve its position within higher education. The ability of geographers to make the world more understandable constitutes *progress in geography*. This paper has reviewed the current asymmetrical contexts within which such progress is achieved in Anglo-American and African academic circles, and illustrated how professional interaction is implicated in all three types of geographical progress. Furthermore, the paper posits (a 'new') interaction as an integral part of the pursuit of geographical progress, and of the continual remaking of Anglo-(American) African(ist) academic geography. It thereby implicates uncritical interaction in the perpetuation of the asymmetrical relations that govern the production of academic geography.

Relatedly, the paper illustrates how interaction, 'new' or otherwise, consists of flows and movements of material and intellectual resources, which create networks of power/knowledge, 'new contexts of action and new arenas of self-presentation and reaction to others' (Dodds 1994: 88; see also King 1990). And, as these, like all forms of power/knowledge, serve a variety of interests, and have their quite specific material geographies of (inter)national and (intra)disciplinary domination and rebellion, these geographies have also been the subject of critical review. However, as considerably more professional interaction between Anglo-American and African geographers takes place within networks of solidarity/trust than within the context of large formal institutional linkage schemes, this creates space for geographers to exercise considerable personal discretion in a morally defensible manner, albeit within limits imposed by structural constraints (Smith 1994; Wilbanks 1993). In this connection, the enormously complex and frequently contradictory dynamic of asymmetrical relations between African and Anglo-American geographers poses real ethical dilemmas for critical geographers.

Does professional interaction sufficiently challenge the processes which contribute to the marginalisation of 'overseas geographies' within mainstream Anglo-American human geographies? Does it directly engage with the forces that lead to the relative isolation of the African academic system (and with it African geographies and geographers) within global academe? Would a 'new' interaction merely facilitate the continued psychological and economic affirmation of the unequal power relations linking the Anglo-American and African academic systems? Can there be more than mere pretence to full/equal partnerships between rich(er) and poor(er) geographers, even within the context of a 'new' and progressive interaction? What sort of power and rights does interaction bestow on Anglo-American geographers and geographies? What opportunities for self-(re)presentation does a 'new' interaction offer Anglo-American geographies and geographers? Is the unquestioned acceptance of, or even reluctant resignation to, the geographical articulations represented by the 'unequal exchange' of North–South interaction tantamount to a conspiracy of silence? Does such silence implicate and compromise geographers in the perpetuation of poverty, debt and inequality (*cf.* Madge 1993)? What are the obligations on African geographers to repay 'debts of gratitude' incurred as part of professional interaction?

These are questions that demand very careful contemplation. For, and the quote from David Smith which opens this conclusion reminds us forcefully of this, 'new' interaction thinking is embedded in wider discourses of how 'geography [is] saturated in politics' (Dodds 1994: 95); why geography is about power (Driver 1994); and when the ethical questions raised by the way we treat distant 'other' scholars may be construed as part of a wider political commitment (Smith 1994). Furthermore, interest in intra-community relations in geography fits in with current debates about the relationship between European imperialism and geography and about contemporary academic imperialism and non-European regions of the world (Berg and Kearns 1998; Dodds 1994; Kinyanjui 1994; Madge 1993; McGinn 1994; Sidaway 1993; 1997b). Consequently, the contradictory and, sometimes, conflictual context of professional interaction between geographers is, arguably, as important as either the precise forms that such interaction actually takes, or the reactions that it generates.

Acknowledgements

I am grateful to James Sidaway and Clare Madge for endless discussions and comments, and to Tony O'Connor, Bill Adams, Lovemore Zinyama, Peg Peil, Paulo Farias, Katie Willis, Elsbeth Robson, Hamish Main, Karin Barber, Simon Batterbury, David Simon, Jane Guyer, Frank Jegede, Gina Porter, Rick Schroeder, A.B. Mamman and Jenny Elliott for commenting on various drafts. I acknowledge the encouragement of John Agnew and Paul Richards, and recognise the input of Musa Ilya, Kojo Amanor, Ken Swindell, Mike Mortimore and other colleagues with whom I have debated these and related issues, but who have not had the opportunity to comment directly on the paper. I alone remain responsible for the ideas expressed; the specific uses to which I have put comments, information and confidences of generous colleagues; and an eclecticism that probably borders on the criminal in its perverse insistence on extending the context and meaning of (other people's) work which provided inspiration.

References

Abler, R.F. (1988) Awards, rewards and excellence: keeping geography alive and well, *Professional Geographer* **40**: 135–140.

Abler, R.F. and **Baerwald, T.J.** (1989) How to plunge into the research funding pool, *Professional Geographer* **41**: 6–10.

Adedeji, A. (ed.) (1994) *Africa Within the World: Beyond Dispossession and Dependence.* Zed Books, London.

Adepoju, A. (ed.) (1993) *The Impact of Structural Adjustment on the Population of Africa.* United Nations Population Fund in association with James Currey and Heinemann, London.

Anon. (1993) Ajayi's education plan. *West Africa* **3976** (6–12 December): 2211.

Baker, J. and **Pedersen, P.O.** (eds) (1992) *The Rural–Urban Interface in Africa. Expansion and Adaptation.* Scandinavian Institute of African Studies, Uppsala.

Bako, S. (1994) Education and adjustment in Nigeria: conditionality and resistance. In Diouf, M. and Mamdani, M. (eds) *Academic Freedom in Africa,* 150–175.

Bangura, Y. (1994) Intellectuals, economic reform and social change: constraints and opportunities in the formation of a Nigerian technocracy *Development and Change* **25**: 261–305.

Barbour, K.M. (1982) Africa and the development of geography, *Geographical Journal* **148**: 317–326.

Barker, D. (1986) Core, periphery and focus for geography? *Area* **18**: 157–160.

Batterbury, S. (1997) Alternative affiliations and the personal politics of overseas research: some reflections. In Robson, E. and Willis, K. (eds) *Postgraduate Fieldwork in Developing Areas: A Rough Guide* (2nd edn), 85–112.

Batty, M. (1994) Editorial: the job market for PhDs, *Environment and Planning B: Planning and Design* **21**: 257–8.

Bebbington, A. (1997) Social capital and rural intensification: local organisations and islands of sustainability in rural Andes, *The Geographical Journal* **163**: 189–197.

Berg, L.D. and **Kearns, R.A.** (1998) America Unlimited, *Environment and Planning D: Society and Space* **16**: 128–132.

Bradshaw, M.J. (1990) New regional geography, foreign-area studies and Perestroika, *Area* **22**: 315–322.

Cliffe, L. and **Seddon, D.** (1991) Africa in a New World Order, *Review of African Political Economy* **50**: 3–11.

Cline-Cole, R. (1996) African and Africanist biodiversity research in a neo-liberal context *Africa* **66**: 145–158.

CODESRIA (Council for the Development of Social Science Research in Africa) (1996) *The State of Academic Freedom in Africa 1995.* CODESRIA, Dakar, Senegal.

Cohen, S.B. (1990) Geographical gender studies. Fresh approaches but with integration, *Professional Geographer* **42**: 231–232.

Corbridge, S. (1993) Marxisms, modernities and moralities: development praxis and the claims of distant strangers, *Environment and Planning D: Society and Space* **11**: 449–472.

Crittenden, R. (1988) Breaking the rules: geography goes 'gung-ho', *Area* **20**: 372–373.

Crowder, M. (1987) 'Us' and 'them': the International African Institute and the current crisis of identity in African studies, *Africa* **57**: 109–122.

Crush, J.S., Reitsma, H. and **Rogerson, C.M.** (1982) Decolonising the human geography of Southern Africa, *Tijdschrift voor Economische en Sociale Geografie* **73**: 197–198.

Currey, J. (1986) The state of African studies publishing, *African Affairs* **85**: 609–612.

Dando, W. (1989) Grantsmanship, *Professional Geographer* **41**: 3–6.

Dawson, J. and **Hebden, R.** (1984) Beyond 1984 – the image of geography, *Area* **16**: 254–256.

Dickenson, J.P. (1986) Irreverence and another new geography, *Area* **18**: 52–54.

Diouf, M. and **Mamdani, M.** (eds) (1994) *Academic Freedom in Africa.* CODESRIA, Dakar, Senegal.

Dodds, K-J. (1994) Eugenics, fantasies of empire and inverted Whiggism. An essay on the political geography of Vaughan Cornish, *Political Geography* **13**: 85–99.

Driver, F. (1992) Geography's empire: histories of geographical knowledge, *Environment and Planning D: Society and Space* **10**: 23–40.

EEC (European Economic Community) (1991) *Life Sciences and Technologies for Developing Countries. 1991–1994 Work Programme.* Commission of the European Communities, Brussels.

ESRC (Economic and Social Research Council) (nd) *Global Environmental Change Programme.* ESRC, Swindon.

Farmer, B.H. (1983) British geographers overseas, 1933–1983, *Transactions of the Institute of British Geographers, New Series* **8**: 70–79.

Gana, A.T. (1993) The Nigerian university at the cross road, *Annals of the Social Science Council of Nigeria* **5**: 1–15.

Gefu, J. (1992) Part-time farming as a survival strategy: a Nigerian case study. In Baker, J. and Pedersen, P.O. (eds) *The Rural–Urban Interface in Africa,* 295–302.

Gibbon, P., Bangura, Y. and **Ofstad, A.** (eds) (1992) *Authoritarianism, Democracy and Adjustment: the Politics of Economic Liberalisation in Africa.* Scandinavian Institute of African Studies, Uppsala.

Gilbert, A. (1987) Research policy and review 15. From Little Englanders to Big Englanders: thoughts on the relevance of relevant research, *Environment and Planning A* **19**: 143–151.

Hagan, G.P. (1994) Academic freedom and national responsibility in an African state: Ghana. In Diouf, M. and Mamdani, M. (eds) *Academic Freedom in Africa*, 39–72.

Haggett, P. (1990) *The Geographer's Art.* Basil Blackwell, Oxford.

Haigh, M.J. (1982) The crisis in American geography, *Area* **14**: 185–189.

Hargreaves, J.D. (1985) British policy and African universities: Sierra Leone revisited, *African Affairs* **84**: 323–330.

Harrison, M.E. (1995) Images of the Third World: teaching a geography of the Third World, *Journal of Geography in Higher Education* **19**: 285–297.

Himmelstrand, U., Kinyanjui, K. and **Mburugu, E.** (eds) (1994) *African Perspectives on Development.* James Currey, London, 280–295.

Howes, D.W. (1989) Sudanese geography: recent research at the University of Khartoum, *Professional Geographer* **41**: 214–217.

Hyden, G. (1996) African studies in the mid-1990s: between Afro-pessimism and Amero-skepticism, *African Studies Review* **39**: 1–17.

Iloegbunam, C. (1993) In deep crisis, *West Africa* **3977** (13–19 December): 2248.

Imam, A. and **Mama, A.** (1994) The role of academics in limiting and expanding academic freedom. In Diouf, M. and Mamdani, M. (eds) *Academic Freedom in Africa*, 73–108.

ISSER/DPPC (Institute of Statistical, Social and Economic Research, University of Ghana, Legon and Development and Project Planning Centre, University of Bradford, UK) (1998) *Women in Public Life in Ghana.* ISSER, Accra, Ghana.

Johnson, M. (1989) Expatriates in African studies, *African Affairs* **88**: 77–82.

Johnston, R.J. (1984) The world is our oyster, *Transactions of the Institute of British Geographers, New Series* **9**(4): 443–459.

Johnston, R.J. (1985) To the ends of the Earth. In Johnston, R.J. (ed.) *The Future of Geography*, Methuen, London.

Kay, G. (1993) Whither Africa and development studies? *Area* **25**: 137–141.

Kenzer, M. (1993) Applied academic geography and the remainder of the twentieth century, *Applied Geography* **12**: 207–210.

Kidd, A.D. (1993) Analysis of an approach to developing 'Viable Policy': a case study of a university linkage project, *Land Use Policy*, **10**(1): 16–25.

King, K. (1990) The new politics of international collaboration in educational development: Northern and Southern research in education, *International Journal of Educational Development* **10**: 47–58.

Kinyanjui, K. (1994) African education: dilemmas, challenges and opportunities. In Himmelstrand, U., Kinyanjui, K. and Mburugu, E. (eds) *African Perspectives on African Development*, James Currey, London, 280–295.

Land, T. (1994) Universities seek autonomy, *West Africa* **3991** (28 March–3 April): 546.

Lawler, D.M. (1992) Environmental geography in the IBG: the new environmental research group, *Area* **24**: 309–315.

Leftwich, A. (1983) *Redefining Politics: People, Resources and Power*. Methuen, London.

Logan, I.B. and **Mengisteab, K.** (1993) I.M.F.–World Bank adjustment and structural transformation in sub-Saharan Africa, *Economic Geography* **69**: 3–21.

Lonsdale, R.E. (1986) The decline in foreign-area specialisation in geography doctoral work, *Journal of Geography* **85**: 263–266.

Lowe, M.S. and **Short, J.R.** (1990) Progressive human geography, *Progress in Human Geography* **14**: 1–11.

Lugalla, J.L.P. (1993) Structural adjustment policies and education in Tanzania. In Gibbon, P. (ed.) *Social Change and Economic Reform in Africa*, Scandinavian Institute of African Studies, Uppsala, 184–214.

Madge, C. (1993) Boundary disputes: comments on Sidaway (1992), *Area* **25**: 294–299.

McCracken, J. (1993) African history in British universities: past, present and future, *African Affairs* **92**: 239–253.

McDowell, L. (1990) Sex and power in academia, *Area* **22**: 323–332.

McGinn, N.F. (1994) The impact of supranational organisations on public education, *International Journal of Educational Development* **14**: 289–298.

Mittelman, J.H. (1997) Academic freedom, transformation and reconciliation, *Issue* **xxv**: 45–49.

Mohan, G. (1994) Destruction of the con: geography and the commodification of knowledge, *Area* **26**: 387–390.

Moore, D.S. (1993) Contesting terrain in Zimbabwe's Eastern Highlands: political ecology, ethnography, and peasant resource struggles, *Economic Geography* **69**: 380–401.

Morgan, W.T.W. (1992) Tropical African colonial experience. In Gleave, M.B. (ed.) *Tropical African Development. Geographical Perspectives*, Longman, Harlow, 5–49.

Mortimore, M.J., Olofin, E.A., Cline-Cole, R.A. and **Abdulkadir, A.** (eds) (1987) *Perspectives on Land Administration and Land Development in Northern Nigeria.* Bayero University, Kano, Nigeria.

Mustapha, R.A. (1992) Structural adjustment and multiple modes of livelihood in Nigeria. In Gibbon, P., Bangura, Y. and Ofstad, A. (eds) *Authoritarianism, Democracy and Adjustment,* 188–216.

Mustapha, R.A. (1993) *Society and the Social Sciences in Northern Nigeria: 1962–1993.* Paper to the African Studies Association (USA) Conference, Boston.

Mustapha, R.A. (1996) The state of academic freedom in Nigeria. In CODESRIA, *The State of Academic Freedom in Africa,* 103–120.

Myers, D. (ed.) (1995) *Reinventing the Humanities. International Perspectives.* Australian Scholarly Publishing, Kew, Victoria.

Neave, G. and **van Vught, F.A.** (eds) (1994) *Government and Higher Education Relationships Across Three Continents. The Winds of Change.* Elsevier and Pergamon, Oxford, England.

Okafor, S.I. (1989) Research trends in Nigerian human geography, *Professional Geographer* **41**: 208–214.

ÓTuathail, G. (1992) Political geographers of the past VIII: putting Mackinder in his place: material transformations and myth, *Political Geography* **11**: 100–118.

ÓTuathail, G. and **Agnew, J.** (1992) Geopolitics and discourse: practical geopolitical reasoning in American foreign policy, *Political Geography* **11**: 190–204.

Parnwell, M. (1994) *Register of Current Research Interests of Members of the Developing Areas Research Group, Institute of British Geographers.* IBG-DARG, Hull.

Paul, B.K. (1993) A case for greater interaction between the geographers of developed and developing countries, *Professional Geographer* **45**: 461–465.

Peet, R. and **Watts, M.** (1993) Introduction: development theory and environment in an age of market triumphalism, *Economic Geography* **69**: 227–253.

Peil, M. (1986) Leadership of Anglophone tropical African universities, 1948–1986, *International Journal of Educational Development* **6**: 245–261.

Peil, M. (1995) Academics and African governments: a shaky relationship. In Myers, D. (ed.) *Reinventing the Humanities,* Australian Scholarly Publishing, Kew, Victoria, 96–102.

Porteous, J.D. (1986) Intimate sensing, *Area* **18**: 250–251.

Porteous, J.D. (1988) No excuses, Belinda, *Area* **20**: 72.

Porter, G. (1995) 'Third World' research by 'First World' geographers: an Africanist perspective, *Area* **27**: 139–141.

Potter, R.B. and **Unwin, T.** (1988) Developing areas research in British geography 1982–1987, *Area* **20**: 121–126.

Potter, R.B. (1993) Little England and little geography: reflections on Third World teaching and research, *Area* **25**: 279–291.

Proctor, J.D. (1998) Ethics in geography: giving moral form to the geographical imagination, *Area* **30**: 8–18.

Robinson, D.J. and **Long, B.K.** (1989) Trends in Latin Americanist geography in the United States and Canada, *Professional Geographer* **41**: 304–314.

Robson, E. and **Willis, K.** (eds) (1997) *Postgraduate Fieldwork in Developing Areas: A Rough Guide* (2nd edn). Developing Areas Research Group of the Institute of British Geographers/Royal Geographical Society, London.

Rogerson, C.M. and **Parnell, S.M.** (1989) Fostered by the laager: apartheid human geography in the 1980s, *Area* **21**: 13–26.

Rundstrom, R.A. and **Kenzer, M.S.** (1989) The decline of fieldwork in human geography, *Professional Geographer* **41**: 294–303.

Sada, P. (1982) Geography in Nigeria, perspectives and prospects: a presidential address, *Nigerian Geographical Journal* **26**: 1–13.

Samoff, J. (1990) The politics of privatisation in Tanzania, *International Journal of Educational Development* **10**: 1–16.

Sant, M. (1992) Applied geography and a place for passion, *Applied Geography* **12**: 295–298.

Sawyerr, A. (1994) Ghana: relations between government and universities. In Neave, G. and van Vught, F.A. (eds) *Government and Higher Education Relationships Across Three Continents*, 22–53.

Schroeder, R. (1997) 'Re-claiming' land in the Gambia: gendered property rights and environmental intervention, *Annals of the Association of American Geographers* **87**: 487–508.

Shresta, N.R. and **Davis, Jr, D.** (1989) Minorities in geography: some disturbing facts and policy measures, *Professional Geographer* **41**: 410–421.

Sidaway, J.D. (1990) Post-Fordism, post-modernity and the Third World, *Area* **22**: 301–303.

Sidaway, J.D. (1992) In other worlds: on the politics of research by First World geographers in the Third World, *Area* **24**: 403–408.

Sidaway, J.D. (1993) The decolonisation of development geography? *Area* **25**: 299–300.

Sidaway, J.D. (1997a) The (re)making of the western 'geographical tradition': some missing links, *Area* **29**: 72–80.

Sidaway, J.D. (1997b) The production of British geography, *Transactions of the Institute of British Geographers, New Series* **22**: 488–504.

Simon, D. (1994) Putting South Africa(n geography) back into Africa, *Area* **26**: 296–300.

Simon, D. (1997) Development reconsidered: new directions in development thinking, *Geografiska Annaler* **79B**: 183–201.

Sindiga, I. (1988) Environmental Studies in Kenya, *Professional Geographer* **41**: 492–493.

Sindiga, I. and **Burnett, G.W.** (1988) Geography and development in Kenya, *Professional Geographer* **40**: 232–237.

Slater, D. (1989) Peripheral capitalism and the regional problematic. In Peet, D. and Thrift, N. (eds) *New Models in Geography. Volume Two*, Unwin Hyman, London, 267–294.

Smith, D.M. (1994) On professional responsibility to distant others, *Area* **26**: 359–367.

Smith, S.M. (1986) UGC research ratings: pass or fail? *Area* **18**: 239–245.

SSRC (Social Science Research Council) and **ACLS** (American Council of Learned Societies) (1989) *The Project on African Agriculture. Draft Proposal*. Mimeo.

Sunal, D.W. and **Sunal, C.C.** (1991) Professional and personal effects of the American Fulbright experience in Africa, *African Studies Review* **34**(1): 27–56.

Swindell, K. (1990) International labour migration in Nigeria 1976–1986: employment, nationality and ethnicity, *Migration* **8**: 135–155.

Thrift, N.J. (1985) Taking the rest of the world seriously? The state of British urban and regional research in a time of economic crisis, *Environment and Planning A* **17**(1): 7–24.

Unwin, T. and **Potter, R.B.** (1992) Undergraduate and postgraduate teaching on the geography of the Third World, *Area* **24**(1): 56–62.

Wilbanks, T.J. (1989) Research funded by mission agencies, *Professional Geographer* **41**: 15–19.

Wilbanks, T.J. (1993) President's Column, *AAG Newsletter* **28**: 1–2.

Woods, R. and **White, P.** (1989) Population structure and dynamics of minority groups, *Area* **20**: 201–202.

Wooldridge, S.W. (1950) Reflections on regional geography in teaching and research, *Transactions, Institute of British Geographers* **16**(1): 1–12.

Zinyama, L.M. (1988) Human geography in Zimbabwe: a review of past research and current trends, *Geographical Journal of Zimbabwe* **19**: 59–79.

CHAPTER 6

Do they need ivy in Africa?

Ruminations of an African geographer trained abroad

Daniel S. Tevera

Introduction

It is now twenty years since I enrolled at Queen's University in Canada and eighteen years since I joined the University of Cincinnati in the USA for postgraduate studies in geography. During the past fourteen years I have been teaching geography and doing research almost continuously at the University of Zimbabwe, except for the 1992/93 academic year, which I spent in the department of environmental science at the University of Botswana as a visiting scholar, and other interruptions when I have been away in Sweden, England, the USA and Germany on brief research fellowships. In this chapter, I intend to ruminate on my experiences as a student in postgraduate geography programmes at the two North American universities and as a staff member at the University of Zimbabwe.

The chapter is largely rooted in past and present personal experiences both in North America as a student and as an African academic geographer. The primary goal is to examine why geography in sub-Saharan Africa is facing a crisis which it might not have the resilience to recover from. Several reasons have been advanced to explain the crisis: African geography has uncritically accepted the epistemologies and social constructions of mainstream Anglo-American geographic thought with insufficient effort to contextualise its application, and the moribund economies of many sub-Saharan African countries are seriously threatening scholarship. These economies are symptomatic of the African crisis, which will not be discussed in this contribution as it has been adequately explained by a burgeoning collection of dissimilar viewpoints ranging from modernisation to postcolonialism and post-traditionalism (Nabudere 1997).

Is training acquired abroad relevant?

Many sub-Saharan African countries still rely on foreign degree-awarding educational institutions for the training of their nationals, especially at the postgraduate level and in specialist fields such as medicine and computers. This will continue to be the trend until these countries establish academic institutions that have the capacity to provide the cadre of trained staff needed to facilitate economic growth and national self-reliance. Opportunities for postgraduate and specialist training are restricted by financial and personnel constraints. The training acquired abroad is generally adequate and has often facilitated the adoption of Western technology, which is considered by many to be an advantage. Sceptics of foreign education, especially from culturally different countries with vastly different labour–capital ratios, question whether it adequately prepares African students for developmental roles that they will be called upon to perform in their native countries. In African newspapers, one often reads frustrated remarks by officials that they do not need staff who are well versed in temperate or other developmental techniques but ignorant of the strategies that are best suited to the tropical setting of most Third World countries.

A question asked repeatedly in Zimbabwe is whether foreign academic programmes provide comprehensive training that best prepares African graduates to function adequately when they return to what are radically different situations back home. Human geography is a social science and having studied it abroad I cannot resist the temptation to share my thoughts based on some personal experiences both as a student in North America and as an academic geographer working in Zimbabwe.

Since the training strategies employed in North American universities are tailored to meet the national development needs of a socio-cultural milieu that is radically different to ours, it is not surprising that some of the Africans who have gone through these domestic programmes have failed to apply the acquired knowledge in their home countries. The instructional programmes, which are ideal for local students, often place international students in a straitjacket of specialisation that is restrictive and that limits their economic usefulness back in their native countries. In that respect, some of the North American and European academic programmes could have fallen short in providing comprehensive training that best prepares students to function effectively in their home countries. The question then is, does this mean that universities in the North need to revisit their international training programmes if they are to fulfil their responsibility to international students from developing countries, who are often assigned leadership positions immediately on return to their home countries?

During the late 1970s, I was one of the numerous students from developing countries enrolled in North American universities. The number doing geography at the postgraduate level was quite small, maybe because geography has never been a popular discipline in that part of the world. In fact, several departments were facing the threat of closure for various reasons, including an inability to attract enough students. Geography in North America then attracted few students from minority groups. A study by Shrestha and Davis (1989: 410) revealed that, by the late 1980s, women and minorities (blacks, Hispanics and Native Americans) still constituted 'just a tiny fraction of the total number of student and professional geographers in the United States'.

However, the postgraduate programme in geography at Queen's University in Canada was expanding at the time and was able to attract many international students. At Queen's, I was exposed to a full repertoire of new conceptual approaches, analytical techniques and critical insights not only from the discipline of geography but also from economics, political science and history. The philosophy of my thesis supervisor, Professor Barry Riddell, was that one became a better geographer if one also read extensively from the literatures of related disciplines. Following Professor Riddell's advice, I took courses in these subjects, which I found quite demanding because I had never studied them at university before. I was critical of the approach then but have since appreciated the logic because to understand the broad field of development, which I am interested in, one needs to be interdisciplinary. In fact, there is much merit in the North American approach, which demands that postgraduate students take courses from outside their discipline. At the University of Cincinnati, doctoral students were required to take four courses (twelve credits) outside their departments before they could qualify to write the comprehensive examinations that preceded the thesis-writing stage. I took more than the four required courses from the department of public administration and political science and ended up reading for a concurrent degree in policy analysis, which I also received when I obtained my doctoral degree in geography.

In Canada and the USA, the interaction with postgraduate students from related disciplines and other parts of the world exposed me to new paradigms of scholarly procedure. I broadened my intellectual horizons considerably from the often heated informal discussions that we engaged in following the regular Friday seminars during which visiting academics, teaching and doctoral students presented the findings of their most recent studies. There are instances during moments of nostalgia or disquietude that I contrast, in my mind, the limited opportunities we have at the University of Zimbabwe to engage in intellectual discourse on the methodological and philosophical (epistemological and ontological) issues that affect the discipline because of the need to attend to more practical issues or 'survival strategy' matters, which I discuss later in this chapter, with my postgraduate days, when there was always time for that. During my postgraduate studies, I attended several conferences of the Association of American Geographers (AAG), the Canadian Association of African Geographers and the African Studies Association. Of particular interest were the specialist group meetings, especially those on Africa and development in general.

At the University of Cincinnati, postgraduate students were required to read Sauer, Hartsthorne, Schaefer and Kuhn so that they would have a good grasp of the nature and purpose of geography and the paradigm shifts that have occurred. At the University of Cincinnati we were a group of postgraduate students pursuing very diverse topics and methodological approaches. During this period of post-everything, the Kuhnian science of the structure of scientific revolutions should be quite relevant. Despite the numerous positive aspects about North American geography, there were times when I found some of the approaches quite eclectic and too polemical. There was too much fetishisation of space, and I often had the distinct feeling of traversing an intellectual wasteland leading to a definite conceptual *cul-de sac*. Somehow I managed to make sense out of the abstract papers that I was required to read in order to pass the required examinations.

At Queen's, my initial idea was to undertake research on spatial inequalities and planning in Zimbabwe for my thesis so that I would return home with a toolkit full of solutions to tackle the problem of regional disparities in my country. My former head of geography at Njala University College (University of Sierra Leone), Dr Harry Turay, now principal of Njala University College, had always emphasised the applied aspects of the discipline and had little time for grand theories. However, after taking a course with Professor Barry Riddell on the geography of Third World underdevelopment and attending a seminar on underdevelopment in Kenya by Professor Colin Leys, I decided to shift my research and not to situate it within the restrictive boundaries of the geographical tradition at Njala but to contextualise it within the dependency theory framework. Then dependency theories had considerable intellectual appeal as a powerful critique of the largely discredited modernisation approach, which had dominated development geography during the 1960s and early 1970s (Amin 1976; Frank 1978). At the time, postmodernism, postcolonialism and post-traditionalism had not yet surfaced to challenge dependency theory and its grand explanations as simplistic analyses of situations that could best be explained by focusing on local realities, cultures and individual factors, as has been advocated in the recent works of Crush (1995), Närman (1997), Simon (1997 and Chapter 2, this volume) and Nabudere (1997).

Eventually, I wrote a thesis about the evolution of a colonial space economy in Zimbabwe with particular reference to capitalist cores and labour reserves. The thesis attempted to examine the forces that triggered and perpetuated labour migration in pre-independence Zimbabwe. Without a research grant, it was impossible to travel to Zimbabwe for field research, so I had to rely on secondary data, which proved quite frustrating because some of the required information was not readily available. The challenge was to analyse the data within the dependency theory framework.

My intention had been to use the master's degree study as a platform for a more theoretically rigorous PhD thesis in the same department, but as fate had it I ended up pursuing an entirely different research agenda, at the University of Cincinnati, on the locational aspects of manufacturing industries in Zimbabwe. The geography department at the University of Cincinnati had established itself as one of the good places to go for urban economic research. Some of the geography faculty at the University of Cincinnati, like the late Professor McNee and Professors Stafford and Selya (my supervisor), had for long been researching locational aspects of private and public sector enterprises.

In North America, I experienced the academic challenges facing all postgraduate students, such as how to get started on a research project, maintaining focus and interest in the research, and how to progress at the rapid pace required by the supervisor, who had to spend some of his time away on a different continent doing research. The period of thesis writing is a difficult one, and I remember occasionally sequestering myself voluntarily for several days in order to have sufficient time to work on my thesis before my assistantship ran out.

The other challenges emanated from being a foreign student and a teaching assistant (TA). Like most foreign students, I had to cope which the challenge of pursuing higher education in an environment with a different value system, intellectual traditions and pedagogical style. Also, we had to cope with the prevailing intellectual elitism, which had so much faith in the grand explanations rather than in the unique experiences of individuals, and whose constructed images of the less developed countries were often inconsistent with the reality. As a TA, which is how many postgraduate students earned money to take them through graduate school, I felt thrown into the deep end, having attended no workshops to prepare me for the role.

The demands on TAs were considerable as we had to assist undergraduate students in a wide range of geography courses, some of which we had done as undergraduates using different approaches and books. But since we had to support ourselves financially, we had to do the work well because those TAs who got negative evaluations risked losing their teaching assistantships. There were times, though, when several of the TAs from the Third World felt that the student evaluations were subjective and were based on the negative stereotypical views about us held by some of the students. However, when I reflect now on my five years' experience in North America as a TA, I am convinced that it provided vital training to which every postgraduate student intending to pursue a career in academia ought to be exposed.

Eurocentric human geography at African universities?

Most African universities have geography programmes that offer undergraduate and postgraduate courses (usually up to the master's level) and commonly have at least one foreigner trained abroad on the teaching staff. Although the size of the teaching staff varies from university to university, four to twenty-two is the typical range of establishment, with eight to fourteen being the average size. In addition to human and physical geography, some departments also offer environmental courses (e.g. the geography department at the University of Zimbabwe, which offers a master's course in environmental policy and planning, and the environmental science department at the University of Botswana, which has a master's degree programme in environmental planning), and land-use planning courses (e.g. the geography department at the University of Dar es Salaam) (McKim 1988; Mashalla 1988).

Geography is a well-established subject at high-school level in African countries, with over 80 per cent of the students taking it from forms 1 to 4, which are the first four years of secondary schooling. Geography graduates normally find it easy to secure employment, and most are employed as geography teachers, land-use and regional planners, and marketing specialists in the private sector. A number of the more established African geographers have penetrated the bureaucratic and political world, and several occupy executive positions in NGOs, government agencies and the private sector.

Until the 1970s, Western Europe was the primary destination for Africans seeking opportunities for higher education, but by the 1980s North America, especially the USA had taken over. Like many of these students, I returned home during the mid-1980s to contribute to the development of my country as a university lecturer. I found the conditions at the University of Zimbabwe (UZ) different to those prevailing in North America in at least three ways. First, at UZ, different pedagogical approaches were used whereby the lecture is the dominant method through which teaching takes place. Undergraduate geography students at UZ are used to having as many as a dozen textbooks for a single course, whereas their American counterparts would be uncomfortable with such an arrangement and generally would prefer a single main text, which the lecturer is supposed to follow closely and which is supplemented by several journal articles. The reason for using multiple texts, apart from exposing students to various perspectives, is that the books commonly used are not designed specifically for the local courses, since they are published abroad and their coverage of the issues tends to be partial. Second, one is required to teach on a wider range of courses, including at least one practical course selected from topics such as statistics, air-photo interpretation, map work, surveying, remote sensing and GIS. Third, one does not have TAs to assist with the marking of course assignments. I clearly recall from my North American experience that it was normal for professors to ask their TAs to teach some of their undergraduate courses on a regular basis.

Table 6.1 shows that the majority of geography academic staff in post at universities in English-speaking countries in southern Africa during the 1995–1997 academic years obtained their highest educational qualifications from Europe and North America (especially the UK, the USA and Canada). Some differences were noted, however, within the region. In Namibia, 100 per cent of the geography staff obtained their highest qualifications from Germany, the original colonial power. Also, all the geography staff in Swaziland and Botswana had obtained their highest qualifications abroad, mostly North America and Europe. The percentages for geography staff who obtained their highest qualifications in Africa range from as low as 20 per cent in Lesotho to 69.9 per cent in South Africa. The long period of relative isolation during the apartheid years up to 1994, when a majority rule government came to power, along with the long history and high quality of some institutions, accounts for the very high percentage of staff who acquired their highest qualifications from local universities. It has been argued that the predominance of academic staff at African universities with foreign qualifications accounts for the course programmes and syllabi, which often mirror those at North American and European universities. However, that is a simplistic analysis because the

Table 6.1 Region where geography academic staff at universities in southern Africa obtained their highest educational qualifications (percentage)

	Home country	Elsewhere in southern Africa	Elsewhere in Africa	Europe	North America	Elsewhere in the world
Botswana	–	–	–	58.8	29.4	11.8
Lesotho	10.0	–	10.0	60.0	20.0	–
Malawi	33.3	–	–	22.2	22.2	22.2
Namibia	–	–	–	100.0	–	–
South Africa	69.9	1.0	1.0	17.4	9.7	1.0
Swaziland	–	–	–	37.5	62.5	–
Tanzania	30.0	–	–	50.0	20.0	–
Zambia	18.2	–	–	54.5	27.3	–
Zimbabwe	27.3	–	–	45.4	27.3	–

Source: Commonwealth Universities Yearbook 1995–96 and 1996–97.

similarities between local and foreign programmes is the result of a wide range of factors.

Contemporary human geography in sub-Saharan Africa is modelled along the lines of Anglo-American geography (Akhtar 1988; Okafor 1989; Rogerson and Parnell 1989; Zinyama 1988) and still shows the imprints of 'past academic colonisation' (Crush *et al.* 1982; Wellings and McCarthy 1983). The philosophical underpinnings, methodology and content are derived from Anglo-American geography through two main processes. First, the foreign books and journals used in African universities provide the mainstay of the academic literature in African universities. These sources do not always capture the socio-cultural nuances or unmask the realities that help to explain some of the practices that hinder development at the local level. Second, the existing linkages between Africa and the North, for example, in the form of scholarships for study abroad and university links involving staff exchanges, maintain this orientation. Rogerson and Parnell (1989) pointed some critical fingers at what they described as the subordination of South African human geography to the dominance of Anglo-American geography. Interaction with Anglo-American geographers is generally stronger than with fellow African geographers elsewhere on the continent because of general communication problems and the high costs involved. Simon (1994) has also shown the persistence of these attitudes and links into the 1990s. Furthermore, it is still much easier, quicker and cheaper to travel from Harare (Zimbabwe) to London (UK) than from Harare to Dakar (Senegal), which is about half the distance, because of the entrenched colonial linkages and the weak ties between different regions of the continent, for example between Francophone West Africa and the rest of sub-Saharan Africa.

Similarly, research has been influenced by research trends abroad and by local conditions. The latter have influenced research through the allocation of research funds. Research projects that are seen as addressing national developmental goals are given higher priority than those that are 'too academic' when allocating funds. On the other hand, external forces, which are not always neutral (Sidaway 1992), have also influenced the content and methodology of local research in two main ways. First, to publish in the highly competitive international journals abroad, African geographers have to write articles that are considered to be pertinent by the editors of the foreign journals. Second, given the general shortage of local funds for research, many African scholars have sourced funding from overseas grant-awarding agencies or have participated in collaborative research with partners from the North, who often define the research parameters.

African human geographers have the responsibility to contribute to development by training future personnel and analysing the various interactions in geographic space with a view to promoting the aspects that result in the betterment of human welfare. As a result, research has been dominated by three broad research thrusts: the people–environment problematique, infrastructure-oriented studies and regional development studies. Since the late 1980s, there has been an expansion in feminist geographical research, mostly focusing on gender issues. The publication, *Changing Gender Relations in Southern Africa*, edited by Larsson *et al.*, is an example of some of the good work on gender issues that has emerged from southern Africa recently. Research topics generally reflect a deep awareness of and active involvement with the changing trends in geography methodology in the world.

At the University of Zimbabwe, which has been instructed by the government to jettison its ivory tower image (Zinyama 1988) and its entrenched intellectual elitism, research projects relevant to nation building usually obtain funding, while those primarily produced for consumption by other professional geographers find it difficult to attract government resources. During the decade following the attainment of Zimbabwean independence in 1980, scholarly geographical research has included the following themes: the subsistence farming sector, rural development, post-independence changes in the urban housing sector, locational patterns of manufacturing industry, water resources and recreational patterns (*ibid.*). During the 1990s, the research areas have broadened to include new themes such as waste management, HIV/AIDS, urban agriculture, urban transport, urban poverty, tourism and sustainable development, cross-border migration, peri-urban agriculture and the various effects of the economic structural adjustment programme (ESAP).

The crisis at African universities

When I joined the department of geography, I was told that there were three areas in which I would be assessed for tenure and promotion: research, teaching and university service. For tenure, one has to perform satisfactorily in all three categories, while for promotion, assessment at the level of outstanding is required in one of the categories, usually research, if one is applying for promotion to the senior lecturer grade.

However, for promotion to the associate professorship and professorship grades, one is required to be outstanding in the research category. This is very much the situation across the continent, although some universities do not attach much significance to teaching and university service when considering individuals for promotion. The crisis facing most African universities has made it difficult for one to pursue scholarly endeavours resulting in promotion to the higher academic ranks, and this has been very frustrating.

The general poor performance of African economies since the 1980s has been a contributing factor to the crisis facing African universities. The implementation of structural adjustment programmes across Africa has resulted in the curtailing of public expenditure, including the financing of higher education (Tevera 1995). As a result, many public universities are crumbling and have introduced a number of measures intended to reduce costs. The belt-tightening measures have made working conditions difficult, thereby discouraging many academic staff from aspiring to anything beyond tenure. Serious scholarship has suffered as a result, since academics face numerous problems, including a lack of research funds, mushrooming class sizes, shrinking library and equipment acquisitions, low salaries, and frequent closure of universities, all of which make forward planning of research activities, sabbaticals and contact visits problematic. Crumbling public universities have become symptomatic of the continent's development crisis, as bankrupt governments are increasingly finding it difficult to support higher education adequately. In Zimbabwe, the problem is compounded by the fact that the number of universities has increased from one in the late 1980s to five at the moment, but the number is expected to rise again – to ten by the year 2000.

The result has been a sad decline in standards as field courses and field trips have been dropped from programmes. The week-long field course previously organised by the geography department for second-year students was abolished during the late 1980s due to the large numbers of students that we were now admitting and the lack of funds to meet the high cost of transporting, feeding and accommodating them. Yet these field courses are a core component of geographical training.

Several of the journals published on the continent have ceased to appear regularly, and this places increased pressure on academic staff to publish in overseas journals. The real value of salaries has fallen precipitously, and many full professors on the continent earn less than the equivalent of US$500 per month, which is not enough to meet their basic needs such as accommodation, food, transport, electricity and water bills, education fees for the children, and health-care bills.

Poor teaching conditions, and limited access to recent books, scholarly journals, electronic networks and professional conferences all affect the capacity to develop a vibrant discipline and also seriously demotivate staff. Hyden (1996: 5) observes that under such conditions it is common for faculty and students to take to the streets over 'poor pay or the fact that salaries or stipends are not being paid on time.' Hyden (*ibid.*: 5) further argues that under such working conditions, 'conducting research becomes a luxury that few can afford, with the result that less and less of the knowledge production of Africa comes from the continent'.

These hardships have compelled many first-rate professors to leave for greener pastures at universities elsewhere. As a result, path-breaking research has dwindled. Since the early 1990s, the migration of geography staff has been from West and East Africa largely to the southern African countries of South Africa, Botswana, Swaziland and Zimbabwe, where economic conditions were perceived to be better (see also Chapter 5). Within southern Africa, the migration pattern in recent years has been to Botswana and South Africa. Recently, an increasing number of African academics have left the continent for jobs at North American and European universities. Those who have remained have looked outside the university for additional income. Not surprisingly, it is quite common to hear of stories of university professors who, in the search for income outside the university, operate small businesses in the vibrant informal sector (a few professors own minibuses, which they use to provide public transport in urban areas) or engage in urban agriculture, with the more established ones engaging in petty consulting for local NGOs.

Conclusion

In this chapter, I have attempted to reflect on my personal experiences as a postgraduate student in North America and on the factors affecting the health of academic geography in Africa. I also addressed the issue of the relevance of Western education and whether it produces graduates with the required skills and expertise or whether it produces 'ivory tower' graduates who are well versed in the techniques of the North but are ignorant of those strategies that are suited to the African setting. In order to maximise the value of their educational programmes, universities in the North that provide training to international students, especially from the South, should consider making the training relevant and practical to the needs of the native country of these students. In situations where the writing of a thesis is involved, it is advisable that the student returns to his/her home country to collect data while working under the supervision of a local supervisor who is conversant with conditions on the ground. Such arrangements could be made through formal links with local universities.

I have also argued that scholarly endeavours at African universities are constrained by shrinking research and teaching resources, and heavy teaching loads which leave one with little time for research, and salaries that do not allow one to meet one's basic needs. The fact that academic staff are compelled to expend their energies pursuing parallel survival strategies, rather than reading, reflecting and publishing in scholarly journals, is symptomatic of the crisis. While there are no easy prescriptions for the problems that afflict most African countries and their public universities, there is a need for development geographers in Africa to develop a new research agenda that examines the prevailing development constraints from an African perspective.

References

Akhtar, R. (1988) Geography in Zambia, *Professional Geographer* **40**(1): 100–103.

Amin, S. (1976) *Unequal Development*. Monthly Review Press, New York.

Crush, J.S., Reitsma, H. and **Rogerson, C.** (1982) Decolonising the human geography of Southern Africa, *Tijschrift voor Economische en Sociale Geografie* **73**(4): 197–198.

Crush, J. (ed.) (1995) *Power of Development*. Routledge, London.

Frank, A.G. (1978) *Dependent Accumulation and Underdevelopment*. Macmillan, London.

Hyden, G.F. (1996) African studies in the mid-1990s: between Afro-pessimism and Amero-skepticism, *African Studies Review* **39**(1): 1–17.

Larsson, A., Mapetla, M. and **Schlyter, A.** (eds) (1998) *Changing Gender Relations in Southern Africa: Issues of Urban Life*. Institute of Southern African Studies, Rome.

Mashalla, S.K. (1988) Geography in Tanzania, *Professional Geographer* **40**(2): 229–232.

McKim, W. (1988) Geography in Botswana, *Professional Geographer*, **40**(2): 227–228.

Nabudere, D. (1997) Beyond modernization and development, or why the poor reject development, *Geografiska Annaler* **79**B(4): 203–215.

Närman, A. (1997) Current development thinking, *Geografiska Annaler*, **79**B(4): 217–225.

Okafor, S.I. (1989) Research trend in Nigerian human geography, *Professional Geographer* **41**(2): 208–214.

Rogerson, C.M. and **Parnell, S.M.** (1989) Fostered by the laager: apartheid human geography in the 1980s, *Area* **21**(1): 13–26.

Shrestha, N. and **Davis, Jr, D.** (1989) Minorities in geography: some disturbing facts and policy measures, *Professional Geographer* **41**(4): 410–421.

Sidaway, J.D. (1992) In other worlds: on the politics of research by first world geographers in the third world, *Area* **24**(4): 403–408.

Simon, D. (1994) Putting South Africa(n geography) back into Africa, *Area* **26**(3): 296–300.

Simon, D. (1997) Development reconsidered; new directions in development thinking, *Geografiska Annaler*, **79**B(4), 183–201.

Sindiga, I. and **Burnett, G.W.** (1988) Geography and development in Kenya, *Professional Geographer* **40**(2): 232–237.

Tevera, D.S. (1995) The medicine that might kill the patient. In Simon, D., Van Spengen, W., Dixon, C. and Närman, A. (eds) *Structurally Adjusted Africa: Poverty, Debt and Basic Needs*, Pluto, London, 79–90.

Wellings, P. and **McCarthy, J.** (1983) Whither southern African human geography? *Area* **15**(4): 337–345.

Zinyama, L.M. (1988) Geography in Zimbabwe, *Professional Geographer* **40**(2): 223–227.

Section 2 Reconstructing development assistance and development co-operation

CHAPTER 7

Getting towards the beginning of the end for traditional development aid

Major trends in development thinking and its practical application over the last fifty years

Anders Närman

Introduction

When Truman institutionalised development assistance, he could hardly have imagined that it would still exist in basically the same form fifty years later. At the same time, we have all become so used to its existence that the mere notion of aid being phased out seems rather unrealistic. The tragedy of aid is that it is still there after fifty years, and the need for it is probably greater than ever before, at least in certain parts of the world. If the purpose of aid has ever been to contribute resources from the North to support local mobilisation towards the establishment of self-sustained development in the South it has failed miserably. As will be elaborated on below, many countries in the South, particularly in Africa, are now more dependent on aid than at any time previously. On the other hand, the beauty of it all is that there is still a lot of well-intended humanitarianism in the North trying to channel its solidarity to the less fortunate in the South. However, in the institutionalised aid structures this has been seriously watered down with numerous other and less noble objectives:

> the international aid process became a seething pot-pourri of humanitarianism, commercial self-interest, strategic calculation and bad conscience – a perfect recipe for all the contradictions, confusion and pathological disorder with which aid-giving is afflicted to this day.
>
> (Hancock 1991: 72)

To add to this confusion, it would be difficult to find any clear guidance from development research on how to proceed along the road to more effective aid practice. In previous chapters, we have seen how the swing of the pendulum in the development debate moves from one extreme to the other. Not only have grand theories of development lost their grandeur, but 'development theory' is also being replaced by 'development thinking'. In his contribution, David Simon (Chapter 2) explores the postmodern tendencies of development thinking as a response to the dominant belief in modernisation not only as a way out of underdevelopment but even as a definition of development *per se*. For any conventional development agency, it must have been far easier to handle modernisation than to grapple with postmodernism. However, it is clear that a *critical* analysis of current trends in development thinking, as referred to, could contribute to improved development practice.

In Chapters 3–6, another dilemma for development thinking is illustrated. The perception of development among actors in research, as well as practice, is shaped in a confrontation between numerous 'irrationalities', such as culture, tradition and ideology. Furthermore, as pointed out in Chapter 1, what is said in the debate is not always as important as what is not said. Behind the official developmental policies there might be a number of concealed considerations, which could be kept hidden for well-intended or devious reasons.

This chapter is not intended as a commemoration of fifty years of development. Nor is it an attempt to write its obituary, in the way that Sachs (1992) wanted to do in relation to the whole concept of development. The ambition is merely to give a reflective analysis of some trends in the development of aid, related to various broader standpoints in development discourse. Key issues to be discussed will be:

- What have been the basic issues concerning development aid during the last five decades?
- To what extent is development aid influenced by current academic development thinking?
- How does the aid community react towards contradictions in the development discourse?
- Can we see any new trends emerging during the late 1990s as a response to experiences gained over the years?

To answer these questions, the chapter is divided into three main parts, followed by some concluding remarks. The starting point in each section is a brief account of the theoretical debate related to that specific issue. In the first part, the basic assumption is that modernisation is a key element in development practice, i.e. how to develop a modern welfare society in the Third World as efficiently as possible. Thereafter, the second part introduces the concept of an actor analysis, studying who are the donors and recipients, plus what motivates their respective agendas. Finally, new trends in aid will be seen in the light of alternative development ideas, under the dominance of a neoliberal development ideology.

International trends in aid will be compared with the specific example of Sweden. In this context, we will not elaborate on certain techniques to deliver aid, but rather look at aid as a phenomenon on its own, focusing on the relevant issues in the aid debate over time. In some instances, it will be possible to explore how the discussion has changed more in semantic terms – how words are used – than in a functional manner. So, for example, aid, assistance and co-operation basically stand for the same thing, even if each term can be used with specific normative connotations. To simplify matters, 'aid' is the term I generally use here, even if there is also an attempt to illustrate the trend of the time in the use of words. One final delimitation has to be stressed, namely that this chapter will concentrate almost exclusively on bilateral aid.

Development aid as a tool for modernisation

Modernisation as a concept

Modernisation has often been regarded as a rather uncritical understanding of what development is. Its principal thesis is that Western modernity is the objective to achieve on a global level. The idea of modernisation can be seen as a logical extension of economic growth theories, adding something of a humane social dimension. Two hundred years ago, Adam Smith developed his ideas on how to pursue the wealth of nations based on economic growth and a free-market economy. Industrialisation came to play a central role in national modernist projects (Preston 1996). To Rostow (1960), an expanding industrial sector was the prime vehicle behind the stages of growth – the ultimate development process.

Later on, the total dominance of the free market, as the only means of achieving growth, was adjusted to allow for a certain measure of authoritative intervention. National development could be encouraged through planning machinery. In addition to this, development assistance was established to boost economic growth and modernisation. During the Cold War, in an attempt to contain the spread of communism, Western material welfare was exposed as superior to any other lifestyle (Preston 1996).

Modernisation might have had its attractions to the poor nations of the world, but there were at least two drawbacks. One was that modernity, as a reflection of Western superiority, lacked something in respect of non-European cultures and traditions. As will be obvious below, traditional cultures and beliefs were often seen as a barrier to development. Additionally, the process of modernisation leads almost automatically to the exclusion of some. Dube (1988: 114) claims that modernisation, as such, legitimises inequity. To Soja (1989: 75), empirical descriptions of modernisation might provide scientific understanding, but they disguise the political meaning. As the spontaneous 'trickle-down' effect of modernisation is, to say the least, grossly exaggerated, it seems to be adequate for politicians adopting the modernisation paradigm if an affluent elite is created. Equity is at best a secondary consideration.

Development aid and modernisation

Since the very initiation of aid, the idea of helping the poor to modernise has been a leading underlying conviction. This has often been motivated by a strong moral commitment to extend the welfare structures of the North to the less fortunate. This section will deal with attempts to extend material welfare in the South, and the difficulty of achieving this noble aspiration. Aid in the context of modernisation will primarily be a matter of how to make the allocated funds function as efficiently as possible.

In 1972, the Development Assistance Committee (DAC) within the

OECD defined aid (Andersson 1989: 60) in terms of three criteria:

1. It has to be handled by official agencies.
2. It has to have economic and material welfare as its main objective.
3. It has to involve a 'grant element' of at least 25 per cent.

Using this definition strictly would rule out not only military aid but also what is given by private voluntary organisations (Cassen and Associates 1986: 2). In the early 1970s, the UN had decided that countries in the industrialised world should allocate at least 0.7 per cent of their GNP for development aid (Andersson 1989: 58).

Development aid is essentially a post-Second World War concept, built on a US initiative by President Truman on bilateral and UN technical assistance (Adams 1993: 62–63; see also Chapters 1 and 2 this volume.). According to Truman, development needed modern technical and scientific knowledge, which was to be shared with underdeveloped countries. As a reason for underdevelopment, it was noted that the economies were stagnant and 'primitive'. At the same time, even if technical assistance was not intended as a new form of exploitation, there were still some benefits to be gained by the donor in terms of expanded economies and international trade (Rist 1997).

In 1949, an international organisation for development assistance was established, namely the Expanded Programme for Technical Assistance (EPTA), which sent out experts and gave scholarships for Third World students. Financing of this body was through voluntary contributions by UN member states (Markensten 1967: 7). Simultaneously, the USA adopted a role as the protector of the 'Free World'. This came a few years after the establishment of the Bretton Woods institutions, which were to safeguard liberal capitalist global policies (Preston, 1996: 168). The scene was set for an institutional structure to promote not only material welfare and socio-economic growth but also the universality in Western values.

Already at the end of the 1960s, many donors had become sceptical about the potentials of aid, leading to a decrease in allocations. In particular, the results of its disbursements were disenchanting the USA. A report commissioned by the World Bank found that aid was important in the transfer of technology, but no strong correlation was found between aid and economic growth (Pearson 1969: 49–52). Possibly, it was assumed, the objectives put too much emphasis on what could actually be achieved in the short run. Both donors and recipients tended to expect too much and too soon, e.g. rapid movement towards an industrial revolution. It was realised that to reach the technological age of the North would take time, and aid had to be combined with internal efforts. Important effects of aid were described as:

> providing the machinery and equipment for industry, for the building of roads and railways, and the creation of new ports and the telecommunication facilities. In agriculture, it has helped considerably to finance the Green Revolution by making available rapid increases in fertilizer supplies, pumps for irrigation, and pesticides for plant protection.
>
> (*ibid.*: 49)

The main motivating force was that the fight against poverty does not stop at national boundaries and it was, therefore, a moral issue to extend welfare programmes. Furthermore, development aid could contribute to increased international trade and international political stability, which was also of benefit to the donors. The report concluded that it was not impossible to expect self-sustained growth in the developing world by the end of the century (*ibid.*: 3–11). The Jackson report on the UN system also stressed the need to regard aid as a long-term commitment, in which capacity building and efficiency were key concepts. In a further report produced with the economist Jan Tinbergen at the helm, a substantial increase in economic growth was stressed (Andersson 1989: 50–54).

Global responsibility for the crisis during the 1980s was stressed forcefully in a couple of reports from international committees. Brandt (1980, 1983) reviewed the acute economic crisis at the time and suggested some kind of new Marshall aid package, with massive resource transfers. In essence, it was another appeal for humanitarian aid on moral grounds. Another common dilemma, discussed by the Bruntland Commission, was the relationship between environmental concerns and development. Sustainability was coined as an expression to be joined together with economic achievements (World Commission on Environment and Development 1987). In both cases, the development paradigm is basically the mainstream Western one, believing in economic growth as the only way out of the development crisis (Rist 1997). Interestingly enough, even the South Commission (1990), which was set up under the chairmanship of former Tanzanian President Julius Nyerere, follows a very similar line in its presentation of *The Challenge to the South*.

An increasingly wide socio-economic gap between the North and the South, as well as within the South itself, could be seen as one indication that aid did not really function well. In addition to this, the escalating debt crisis was getting more and more out of hand. During the 1980s and 1990s, it reached proportions that seemed impossible to solve in any conventional way. The development policy that came to dominate totally from the 1980s onwards was the neoliberal ideology, translated into structural adjustment programmes. The three reports referred to above might have been what Rist (1997: 171) has described as 'a few humanitarian extras'. Donors outside the direct World Bank/IMF sphere often accepted the economic reform programmes, although adding a social dimension. The basis of the argument, for which Reagan and Thatcher became the main symbols, has been summarised thus:

> that the modern free-market capitalist system is maximally effective in producing and equitably distributing the economic, social, political, and intellectual necessaries of civilised life.

> (Preston 1996: 253)

Unfortunately, it has taken some time and a lot of suffering to understand that things are not as simple as all that.

When reconsidering development assistance during the previous twenty-five years, Riddell (1987) noted that the motives behind aid differ, and include national and commercial self-interest and political goals. In spite of all this, he found that most donor governments claim to give aid primarily on moral grounds. If this were really so, it seems odd that the negative effects of aid are focused on so strongly, in a seemingly one-sided way. According to Cassen and Associates (1986), the aid debate might have been distorted by the fact that only selective glimpses of reality were given. Cassen concluded that the primary project/programme goals, as formulated in the terms of reference, are often fulfilled. Even so, a project might still have a negative impact from a broader developmental point of view. In addition to this, it is pointed out that aid to Africa is less efficient compared with what has been experienced elsewhere.

With a view to modernising the South, development aid was established fifty years ago. The general claim was that this effort was built upon a generally felt moral urge. We in the North would share our technological development with the poor. However, this has not functioned in the way originally perceived, and today's world is still characterised by substantial inequalities between both nations and people. The question, then, is whether this is because the aid has not been efficient, or whether some other factors in the analysis have been neglected. Before going into the second part of this question, we will focus on Swedish development assistance, particularly attempts to establish greater efficiency.

Swedish aid and modernisation – a search for efficiency

Official Swedish development assistance dates from an act of parliament in 1962, when Nämnden för Internationellt Bistånd (NIB) – the Board for International Development Assistance – was formed. Three years later, the Swedish International Development Authority (SIDA) replaced it. This institutionalised form of aid was built on a long tradition of humanitarian work, and a principal actor behind the formation of these government bodies was the future prime minister, Olof Palme.

From the very start and much like the rest of the international donor community, official Swedish aid supported the modernisation process. There was a generally felt moral commitment that the expansion of economic growth and subsequent material welfare could not be restricted by national boundaries. At the same time, if this was to be to the advantage of Swedish industrial and commercial interests, it was not regarded as something negative. Economic underdevelopment was identified as the core problem, but attitudes and traditions contributed to the prevailing state of affairs. Some of the early projects within Swedish official development assistance were in education/training and health/family planning (Kungl. Maj:t. 1962).

In the late 1960s, Markensten (1967) reviewed various projects to which SIDA was contributing. Education (teacher training and vocational education) and health (including family planning) were the two dominant sectors at that time. Furthermore, we can find various attempts to modernise the agricultural sector (integrated rural development) and to improve infrastructural standards. The primary objective was to achieve the maximum development effect, even if it was difficult to define what development was. In a slightly cautious manner, Markensten (*ibid.*: 15–16) notes that what we in Sweden regard as development *might be* different from what people in the recipient country are striving for – which depends on their own cultures and values. However, this was not perceived as a serious problem, since most of the leading politicians in developing countries had a Western education. Consequently, they were already assumed to have adopted our values and conceptual framework.

Against such a simplistic notion of modernisation, Knutsson (1965) introduced the existence of varying realities. New techniques can only be assessed in the socio-economic and cultural totality of which they form a part. In trying to define development, some kind of improvement was suggested, but this only opened up a new dilemma. Ideas on what constitutes an improvement differ according to prevailing values. Thus, development assistance can either improve the technical and social situation within traditional societies or change the whole social structure. In the latter case, it is possible that the donor will break down the traditional society without offering any new form of social community and security to the individuals thereby affected. From this, it follows that the best aid strategy would be to attempt some improvement that builds on the traditional society.

Kihlberg (1971) discussed cultural pluralism and relativity in normative values in an attempt to promote research on modernisation. The establishment of problem-oriented research capacity in the Third World itself was suggested. This could lead to an active and national policy of modernisation, built on a new awareness. A similar approach was adopted in a government paper on development research (Statens Offentliga Utredningar 1972), which led to the establishment of the Swedish Agency for Research Co-operation with Developing Countries (SAREC).

In addition to research, evaluations were supposed to produce knowledge on project impact related to the respective planned outputs (Markensten 1967). This might all be well, but what about the unintended effects, which have rather serious consequences for those involved? It was mentioned above (Cassen and Associates 1986), that even if a certain project does have a positive impact, the broader development outcome might be negative. Jacoby (1971) has illustrated such a dilemma relating to certain ecological effects in the promotion of the Green Revolution.

During the 1980s, development co-operation changed its character considerably. Economic realities, both at home and in the South, coupled with the looming debt crisis, were influential in this. Conditionalities, demanding economic reforms, were soon to become mandatory as part of aid packages. Here SIDA was following World Bank/IMF policies closely, sometimes adding some social aspects – 'giving the reforms a human face'. At the same time, people questioned whether we in Sweden could actually afford to be continuously generous in our development co-operation. In 1983, the issue was not *whether* but *how* the aid allocation was to be cut (or watered down). A close connection to the Swedish commercial sector was more evident (Statens Offentliga Utredningar 1972; Andersson *et al.* 1984). Odén (1984) said that Swedish aid policies conformed to the OECD model in respect of these tendencies. To him, it was desirable that development co-operation be separated from the commercial sector and trade, even if this would mean a decrease in volume.

Efficiency is becoming increasingly central in the debate. Edgren (1984) made the point that with a substantial amount of money allocated as aid, it is necessary to impose conditionalities in order to make it more efficient. At the same time, it seems difficult to measure what efficiency really is. Bigsten (1984) pointed out that some aid objectives are simply impossible to measure quantitatively. Democracy can be seen as a factor restricting our development co-operation – regimes that are *too* undemocratic should get no aid at all. He argued that economic growth in itself would lead to greater independence. Consequently, what can be measured would be economic growth and the distribution of wealth. Balancing these two factors against each other is done simply by an expectation that in due course economic growth will benefit the least affluent by means of the so-called 'trickle-down effect'. In the final analysis, aid efficiency can be measured by simply assessing the national economic growth. If this were to be the norm, Swedish aid is not only coming closer to that of other OECD countries but also to the economic thinking of the World Bank/IMF.

Bigsten came to the conclusion that SIDA should allocate its money to projects for which the socio-economic benefits are highest. He assumed that this is also being done intuitively, based on accumulated experience. However, this might be more complicated than assumed, as is evident from an article published over a decade later (Ehrenpreis 1997). Based on numerous evaluation reports, he argued that it is difficult to find a clear correlation between development co-operation and economic growth. According to Ehrenpreis, the causal relationship between aid and economic growth operates in both directions, and sometimes the effects are contradictory. Like Cassen and Associates (1986), he defined the paradox of development aid to be that it is most efficient where it is least needed, and best when not really needed. This aspect would make the reasoning by Bigsten look much more controversial than he himself probably ever assumed.

On the issue of aid efficiency, Karlström (1996: 69–74), claimed that the assumptions that aid has positive development effects could not be verified from any of the large number of macro-economic studies done for years in many different countries. Further, even if some positive effects can be registered from a development project, the overall results can still be found to be economic stagnation or depreciation. Unintended side effects have been suggested as a cause; this must represent something of a fundamental falsification of the whole theory of development aid as such. Using the proposals put forward by Karlström, it is obvious that evaluations assessing only the impact on certain stated objectives, formulated in the terms of reference, are worth next to nothing from a developmental perspective. In addition to this, it can be stated that even if the theory on aid is not elaborate enough to produce a suitable agenda, this is where more research work has to be done. Somehow we have to devote far more energy and openness to an analysis of the relationship between aid and development in a broader context.

In an analysis of development assistance commissioned by the Ministry for Foreign Affairs, it was noted that efficiency was a concept with different interpretations, depending on the scientific approach. Even in the social sciences, there could be variations between anthropologists, political scientists, ecologists and others, but the macro-economic school was still dominant (Statens Offentliga Utredningar 1994: 28). Development aid authorities are alleged not to be able to distinguish between various perspectives on development. SIDA is criticised for not learning and drawing conclusions from evaluations carried out (*ibid*.: 12). From this, it follows that development could be made to function much better, even if not by purely economic criteria, if it were assessed from a multidisciplinary perspective.

In this context, it seems valuable to make some comparisons with the aid debate in Norway, a country similar to Sweden in many respects. Eriksen (1987) has pointed to the lack of competence in relation to the social effects of aid. Similarly, the great need to integrate aid into a wider foreign policy has been mentioned as a priority need. This was also taken up in the new integrated Norwegian parliamentary policy for the South (Utenriksdepartementet 1992). Cultural aspects of development aid have been emphasised in connection with all the so-called 'white elephants' – huge and costly projects with only limited benefit to the recipient (Eriksen 1989). Another dimension of these same cultural aspects is illustrated by Garbo (1993): aid communities build their own societies while on tours of duty in recipient countries – a kind of small Norway or Scandinavia. Consequently, local cultural influence is limited, which must be an inhibiting factor in the professional situation as well.

Development in conflict

Structuralism and the school(s) of dependency

From a normative point of view, the adherents of structuralism and dependency would strongly reject any development policy leading to inequity. An historical division of the world into a core and periphery is still obvious to this day, and will also be reflected within national development tendencies. With its departure from the harmonic models of development, this thinking has been closely linked to neo-Marxism (Preston 1996). As such, it has been heavily criticised from within the 'Left', as well as from the more bourgeois development researchers and practitioners, on both conceptual and empirical grounds. In retrospect, strategies of delinking have been rejected because there would be no way to implement them. The powerful are unlikely to give up a position of power for the benefit of the less advantaged (Frank 1991: 25).

Sometimes, dependency and modernisation have been seen as poles apart in development discourse. However, it is difficult to find what really distinguishes them in defining the central objectives of development. The important dividing line is that while modernisation seems to neglect inequity and conflicting agendas in the development process, these are crucial in dependency thinking – lending it relevance right up to the present. However, to practitioners the dependency schools introduced an element of disagreement, which must certainly still be difficult to grapple with. According to Dumont (1962), the change in rhetoric from 'help' and 'aid' to 'co-operation' is mere hypocrisy. Dube (1988) dismissed the humanitarian aspect of aid on account of the invisible strings attached. Together with trade, and today also debt, neocolonial structures of exploitation were preserved.

Booth (1993: 49) notes in his contribution to the so-called 'impasse debate' that there has been a widening gap between academic research and development policy and practice (see also Chapter 2). This (self-)critique has been aimed mainly at the radical sections of development research. To escape the impasse, a general call for relevancy in development research is not enough; the appropriate question is to whom should it be relevant (*ibid*.: 67)? Emphasis will be given to actor-oriented analysis that integrates the global, national and local levels. According to Long and Long (1992: 272), research should increase our understanding of both intended and also the unintended consequences of human action.

To illustrate these points, actors in development will be focused on below, e.g. donors and recipients. This will involve not only trends in aid volumes given but also objectives of aid – those both openly accounted for and concealed.

Volumes of aid – general trends

One issue that has long been central to the debate on development assistance is the volume given, more often expressed as a percentage of national GNP. Interestingly enough, the levels of aid, expressed in that way, were much higher in 1960 (0.51 per cent) than today (Table 7.1). These data include both bilateral and multilateral aid. In proportional share, there has been a strong shift among various countries since 1960. The most dramatic one is the way that the USA has gone from being the most important donor to a position in which even its absolute contribution has been overtaken by some other states. Simultaneously, some Scandinavian and other medium-sized donors have taken a 'progressive' position, in terms of both volumes and content.

Officially, aid is supposed to benefit the poorest or least developed countries. Since the 1960s, there has been a gradual proportionate increase of aid to the low-income countries from below 40 per cent to well above 60 per cent since the mid-1980s. Even if this is a positive trend, there are certain distortions in the data. So, for example, a number of main recipient countries such as Egypt have been 'downgraded' from middle-income to low-income status. The declining position of the USA among donors is also reflected in this trend, as the main recipients of American aid have always been comparatively well-off. Among the donors, Portugal, Ireland, Denmark, Norway and Finland place the strongest emphasis on low-income countries, while we find the USA, together with Austria, New Zealand, France and Spain, at the other end of the spectrum (Randel and German 1996, 1997).

Table 7.1 Development assistance as a percentage of donor GNP

Donor	1960	1965	1970	1975	1980	1985	1990	1995
UK	0.56	0.47	0.41	0.39	0.35	0.33	0.27	0.28
France	1.35	0.76	0.66	0.62	0.64	0.79	0.60	0.55
Netherlands	0.31	0.36	0.61	0.75	1.03	0.91	0.92	0.81
Belgium	0.88	0.60	0.46	0.59	0.50	0.55	0.46	0.38
Denmark	0.09	0.13	0.38	0.58	0.74	0.80	0.94	0.96
Sweden	0.05	0.19	0.38	0.82	0.79	0.86	0.91	0.77
Norway	0.11	0.16	0.32	0.66	0.85	1.01	1.17	0.87
Finland	–	0.02	0.06	0.18	0.22	0.40	0.63	0.32
Germany	0.31	0.40	0.32	0.40	0.44	0.47	0.42	0.31
Switzerland	0.04	0.09	0.15	0.19	0.24	0.31	0.32	0.34
Ireland	–	–	–	0.09	0.16	0.24	0.16	0.29
Italy	0.22	0.10	0.16	0.11	0.17	0.31	0.31	0.15
Austria	–	0.11	0.07	0.21	0.23	0.38	0.25	0.33
Japan	0.24	0.27	0.23	0.23	0.32	0.29	0.30	0.28
Australia	0.37	0.53	0.59	0.59	0.48	0.48	0.34	0.36
New Zealand	–	–	0.23	0.52	0.33	0.25	0.23	0.23
USA	0.53	0.58	0.32	0.27	0.27	0.24	0.21	0.10
Canada	0.19	0.19	0.41	0.52	0.43	0.49	0.44	0.38
Total	0.51	0.49	0.34	0.36	0.37	0.35	0.34	0.27

Source: World Bank 1986, 1997a, 1998.

In 1996, twenty countries received more than half of all official development aid; fifteen of these were classified as low-income countries. In absolute terms, we find considerable contributions to China, Egypt and India. Four other major recipients are Indonesia, the Philippines and Thailand, i.e. three of the emerging economic 'success' stories in Pacific Asia, plus Bolivia, which are all classed as lower middle-income countries. Interestingly enough, the second largest recipient overall is Israel – a high-income economy (World Bank 1998).

Major donors – trends and motives

In international comparative terms, in 1960 the USA was the most important donor in both proportionate and absolute terms. The USA contributed over 60 per cent of total OECD aid in the mid-1960s. Two decades later, in 1985, the USA share of total OECD aid was down to 32 per cent, reaching a mere 13 per cent ten years later. The USA is now by far the most insignificant donor in proportionate terms, giving only 0.10 per cent of its GNP. One reason for the early decline in the USA, but also the UK, was simply a general policy of budgetary constraints introduced by Reagan and Thatcher respectively (Hayter and Watson 1985: 12).

With the decrease in aid from the USA, other major economic powers have increased their share of the total. France, the UK, (West) Germany and Japan were all overshadowed by the USA in 1965, with an aggregate 30 per cent, reaching 42 per cent in 1985. In 1995, Japan, France and

Germany each gave higher aid contributions than the USA. Together with the UK, they allocated 57 per cent of all OECD aid in that year. As seen from Table 7.1, the proportionate increase in aid from these countries is due to the US decline, rather than to an increased allocation by any of them.

Numerous interests, in addition to the official humanitarian one, determine aid contributions. The World Bank (1997b: 317) has claimed that some countries have traditionally received a high level of aid, for foreign policy reasons: 'Aid dependency ratios may therefore reveal as much about the interests of donors as they do about the need of recipients.'

Strategic US interest, as part of the Cold War, is indicated in the consistently high share of its aid, and thereby also total OECD aid going to countries such as the then South Vietnam, Laos, Jordan, Egypt and Israel (Pearson 1969: 392–393). For all practical purposes, Israel is too well-off even to be considered for aid under normal circumstances. As mentioned, even in the 1990s, Israel remains a major aid recipient, especially from the USA. Israel's share of 1996 US aid was above one-quarter (World Bank 1998).

During the early stages of development aid, a continued affinity between the former colonial powers and their respective ex-colonies is obvious. This is particularly important to France, which gives huge sums of money to semi-autonomous territories such as Réunion, French Guiana, Guadeloupe and Martinique (Hayter and Watson 1985: 18). In the 1990s, France is still transferring a high proportion of its aid to former colonies such as Côte d'Ivoire, Morocco, Algeria and Congo Republic (Brazzaville), as well as numerous other countries in West, North and Sahelian Africa.

Most of Britain's main aid recipients form part of the (previously British) Commonwealth, including India, Bangladesh, Kenya, Tanzania, Malawi and Uganda (Bendix 1996: 25–26; Curtis 1997). Due to the limited size of its aid, the UK could not have played any significant role, except to keep its Commonwealth partners out of the Soviet sphere. In the early 1990s, conditionalities relating to 'good governance' were introduced. Even so, the UK was reluctant to take stern action on human rights against traditional allies, such as Kenya and Nigeria, as illustrated, for example, in the case of Ken Saro-Wiwa's execution (Cumming 1996).

Commercial interests cannot be forgotten in the selection of development aid partners. Patterson (1997) has referred to aid as foreign state investment (FSI) as a more accurate name. He draws up a pattern of interest that also incorporates aid delivery, which is dominated by three regional blocs, i.e. the USA for the Americas, a German-dominated EU for Europe, and Japan for East Asia. One result of this kind of situation might be that Africa will be delinked, not by choice, but by default (Rugumamu 1997).

Part of such a regional aid focus is reflected in the Japanese interest in Asia, which is the main reason for the position of relatively prosperous countries, such as Indonesia, the Philippines and Thailand, among the twenty largest recipients. Other major recipients of aid from Japan include China, India and Bangladesh (World Bank 1998). Here, as elsewhere, the intention is to build markets through a reduction of poverty and capacity building – not charity but enlightened self-interest (*African Business*, May 1988: 25).

Combined strategic and commercial interests are also reflected in the way that aid patterns have changed dramatically during the 1990s, connected to the collapse of the Soviet Union and changes taking place in Eastern Europe. Countries receiving aid have been divided into Part I, the 'traditional' recipients, as defined collectively by the DAC, and Part II, the so-called transitional economies, which are given aid-like resource flows. The division within the Third World has been taken into account, as countries are able to move from Part I to Part II status. As the total amount of development aid has declined, some donors have been re-allocating their aid disbursements from Part I to Part II economies. In the mid-1990s, the largest increase in aid was directed to Central and Eastern Europe (World Bank 1997b: 305–307). These trends have in no way been exclusive to the five major economic powers.

The UK is a middle-ranking donor in terms of proportionate shares. In fact, there is a gradual decreasing trend from 1960 onwards. However, there has been some interest surrounding British aid in the late 1990s. To a large extent this can be connected to the change of government. During the Thatcher government, the share of British aid dwindled, partly as a result of an overall cut in public spending. In addition to this, with declining industrial performance, aid was used as an important tool for export promotion. As late as 1997, the Conservative government introduced a new initiative for industry to work more closely with the aid sector. Since the Labour government took over in 1997, a new ministry for aid has been established, the Department for International Development, led by a full cabinet minister, Clare Short (Bendix 1996: 25–26; Curtis 1997). Even if the new directions seem to have gained a lot of international interest, it might still be too early to assess the new trends.

Like-minded donors – is there an alternative?

The three Nordic countries, Denmark, Norway and Sweden, together with the Netherlands and Canada, came to be known as a like-minded 'progressive' pressure group on development policies. It has been claimed that the main rationale for these countries to give aid has been humanitarian internationalism, which in turn has been based on domestic social values. Among these countries, Canada has the strongest leaning towards commercial self-interest, even if this factor tended to increase for the whole group after 1975 (Stokke 1989). A commitment to development was reflected not least in the volume of aid given by these relatively small countries.

As indicated in Table 7.1, in 1960 Norway, Denmark and Sweden lagged far behind, giving only between 0.05–0.11 per cent. However, development assistance was still not a government responsibility in Scandinavia at the time. An increase in the proportion of total aid from Scandinavia occurred during the early 1960s. This trend continued during the next decades; by 1995, the Nordic countries (including Finland) allocated 8 per cent of total aid. Denmark, the Netherlands, Norway and Sweden have all reached, or at least touched, the 1 per cent goal for aid that they set themselves (*cf.* the DAC target of 0.7 per cent). In Sweden, this had been formulated as a first stage to be reached in the later 1960s (Kungl. Maj:t. 1968). Sweden and Norway have often had something of a neck-and-neck competition to top the donors' league, a position that Norway reached in 1990, with close to 1.2 per cent of GNP.

With the possible exception of Denmark (Table 7.1), a general decline is evident during the 1990s. The prospect is that Norway will be back at 1 per cent again by the year 2000 (Garbo 1997). A similar directive has been issued in Sweden, even if it cannot be realised as early as the end of the century. More serious is what was mentioned above, namely that Sweden is approaching the OECD model of development aid. The same trend has also been followed by Norway (Eriksen 1987), even if its commercial interests used to be less significant. However, Amland (1993) illustrates how Norway is sliding further away from its own model of aid in the 1990s. Commercial interests are becoming a dominant factor in Norwegian aid deliveries too. Garbo (1993), a former Norwegian ambassador to Tanzania, has elaborated on the role of his nation, and the power it is accumulating in its position as a donor country. Part of this reflects the neoliberal policies being followed and the conditionalities that Norway, like other donors, places on the aid. Irrespective of how these strategies were able to assist poor countries, they contained a high level of commercial self-interest.

Since the early 1970s, the Netherlands has been among the three or four most generous donors within the OECD (see also Chapter 8). It would seem that the strong position of development aid in the Netherlands to a large extent reflects the high-profile development minister, Dr Jan Pronk, who was in charge from 1973 to 1979, and again from 1989. Somehow, this is an indication of how certain individual politicians have been able to exert an influence in this sector. This was also noted for the UK with the change from a Conservative to a Labour government, putting Clare Short in charge of aid. It can thus be seen how important internal politics in the North are relative to other criteria or to negotiations with the South.

The Netherlands has only three programme countries, Bangladesh, India and Pakistan. Other recipients are grouped regionally, such as East Africa – Kenya, Tanzania, Uganda and Rwanda. Large amounts are also given to countries in the Andes, Central America and the former colony of Suriname. A considerable amount is channelled through the NGO sector. One aid policy adopted by the Dutch is to concentrate on themes, such as food security or the reduction of infant mortality. Thereby, the Netherlands could work on certain special fields in almost all countries, which is also indicated by the long list of recipients. Even if this system seems to enhance efficiency it is criticised because it will fragment the aid among many countries, and the actual procedure is rather complicated (Bendix 1996; OECD 1997; Wildeman 1997).

While Cooper and van Themaat (1989: 155) have claimed that there was little commercial interest in Dutch aid, the arms trade has been looked into critically. The Netherlands sells a large amount of armaments to countries such as Egypt, Pakistan and Indonesia (Wildeman 1997). This is a similar critique to that often launched against Sweden.

In conclusion, we can see that these 'progressive' donors have recently come much closer to mainstream aid policies. Volumes are declining even if there are declarations that they will be restored. Content-wise, there is ever-increasing support to commercial sectors. As part of this, we also find the wholehearted endorsement of the economic reform programmes, which might not be to the advantage of the poor in the Third World, but to the donor countries themselves. Below, a more detailed account will be given of Sweden specifically.

Some effects of aid

One major dilemma connected to aid is the dependency created by aid itself. An increasing aid dependency was noticed during the 1980s, particularly in sub-Saharan Africa. In low-income countries, apart from China and India, aid finances close to 50 per cent of total investments (World Bank 1997b: 317). Karlström (1996) has introduced the simple logic that if aid had been successful, aid dependency would have been reduced, but instead it has increased. Connected to the dependency syndrome is the issue of conditionalities – the donors' self-imposed right to connect demands for changes in socio-political systems to mere financial inducements. In addition to this, he brings out the cost to the recipients of actually administering the development assistance obtained. Given limited national resources, the financial burden in this respect is substantial.

In any assessment of development aid, we cannot neglect the role of structural dependencies, individual actors' vested interests and power relations. In a case study from Kenya (Porter *et al.* 1991), it has been possible to explore the outcome of a project as an amalgam of a variety of views and opinions held by numerous stakeholders. Using the example of the Sudan, Morton (1996) suggested a more direct, bottom-up approach to aid, pointing to the lack of government will and accountability. Building further on that, Rugumamu (1997) regards aid as a tool used by donors and recipients in an attempt to exercise their statecraft. Attempts to create real independence and self-reliance will not be helped by traditional development aid.

Some notes on non-OECD donors

What is often forgotten in the discussion of international aid is the fact that there are many donor countries outside the OECD. For a long period of time, Organisation of Petroleum Exporting Countries (OPEC) member countries were substantial actors in international development co-operation. In fact, in the late 1970s, Saudi Arabia was at par with US aid in total dollar terms, giving 5–8 per cent of its GNP annually. Other substantial donors were Kuwait and the United Arab Emirates. In proportionate shares, Iran, Iraq, Libya and Qatar have been well above most OECD countries in various years (World Bank 1986).

Previously, the Soviet Union and East European countries also gave aid, even though they have now been turned into major aid recipients. This was channelled through the Council for Mutual Economic Assistance (CMEA) within COMECON, to which the Soviet Union was the major donor. Among its most prestigious projects was the Aswan Dam in Egypt. Somehow, this can be seen as the other side of the aid-giving competition, between the ideology of socialism and the US concept of a free world (Preston 1996: 168–169). COMECON played a certain role as a lobby on trade issues and also for military aid, not least to liberation movements. The direct inclusion within COMECON of three non-European member states, i.e. Cuba, Mongolia and Vietnam, also gives an indication of the direction in the aid flows (Jensen and Nielsen 1981).

As an example of East European aid we can mention the former German Democratic Republic (GDR), which was estimated to have given 0.68 per cent of GNP just before its merger with West Germany. Consequently, its share was even higher than that of its new partner. The GDR concentrated its aid to sectors such as education and vocational training. At the time of the merger, 95 per cent of its aid was directed to eleven countries – Cuba, Vietnam, Mongolia, Laos, Cambodia, Angola, Mozambique, Ethiopia, South Yemen, Nicaragua, Afghanistan – plus the ANC (South Africa), SWAPO (Namibia) and the PLO (Palestine) (Schleicher 1995: 23–25).

Hayter and Watson (1985: 6) made the point that while the DAC reports aid statistics from India, it does not include two other donors, namely China and Cuba. When China re-entered the international scene after the 'Cultural Revolution' in the early 1970s, it tried to re-establish itself as an important Third World nation and built alliances based on the Non-Aligned Movement. In line with its overall foreign policy, it gave aid to African liberation movements and to allied countries such as Tanzania. The most spectacular example of the latter is the Tanzania–Zambia railway (Yu 1975). South Korea and Taiwan are also emerging donors (Raffer and Singer 1996: 120).

Sweden as an actor in development co-operation

Sweden is generally regarded as a 'progressive' donor. In this section, we will look at who the actors in aid delivery are from a Swedish perspective. This will include an account of the organisational structure and its historical roots. Thereafter, there will be an analysis of which countries have been selected for co-operation. Factors such as country programming and conditionalities will be elaborated on. Finally, two other actors in Swedish aid are dealt with, namely the private commercial sector and NGOs. To a large extent, this section is based on internal annual reports and budget proposals from SIDA/Sida.

When the guidelines for official Swedish aid were drawn up and decided upon by parliament in 1962, it was a way to institutionalise the extensive humanitarian work already being undertaken. This included missionary activities in education and health, as well as aid through organisations such as the Red Cross and Save the Children Fund. Democracy had been promoted in the developing world through trade unions and co-operatives. In addition to this, direct commercial investments and economic policies (including a positive stand on free trade) were regarded as being favourable to Third World interests (Kungl. Maj:t. 1962: 3–4). Early state involvement in aid was channelled through the Swedish Institute, including the recruitment of experts and the administration of scholarships on behalf of the UN. Until the early 1960s, the government was more of a partner to voluntary development organisations.

An official body had been formed in the early 1950s – Central-kommitten för Svenskt Tekniskt Bistånd (CK) – the Central Committee for Swedish Technical Assistance – with representatives from the government, the commercial sector and the NGOs. Among the first projects were a technical institute in Ethiopia and a vocational school in Pakistan. A children's hospital was built in Ethiopia with money collected in a huge voluntary exercise (Markensten 1967: 9). Voluntary work raised awareness of the problems facing people in the Third World. To a large extent, the NGOs of today have lost this role as general opinion formers. This poses a dilemma in relation to present ambitions to achieve a kind of partnership in development (see below).

CK was replaced by NIB in 1962; it was followed in turn by SIDA three years later. Development aid is a complicated issue, balancing demands from the donor government and the recipient. SIDA has often been attacked by the mass media in an attempt to score easy points. The Timbro publishing house has also expressed many critical views. In the translated version of Bauer (1984), for example, development assistance is blamed as the main force behind poverty and stagnation in the Third World, along with the confrontation between North and South.

Therefore, a need to show short-term results often conflicts with the potential to achieve longer-term results. It is not difficult to scandalise limited success after a short period of time, when the objective is to achieve long-term effects. Consequently, evaluations have tended to devote much energy to a symbolic legitimising of the projects (Riksrevisionsverket 1988: 142–146). For continuous mutual capacity building, it might be advantageous if a dialogue could be opened with more serious partners than the tabloid press. That has been stressed by the National Audit Board, which stated that SIDA does not have a coherent opinion on what good development aid actually is (*ibid.*: 151). Various sectors have different ideas about what kind of knowledge is needed within the organisation. SIDA has developed from an organisation built on idealism to a professional administration – a shift that has impacted on the whole concept of development aid. The institutionalisation of humanitarian objectives and solidarity might be contradictory in certain terms.

It would take too long to elaborate on all the organisational changes in the official Swedish aid structure since 1965. Two points are of direct interest in this context. First, the 1970s were the decade during which numerous new bodies were set up, while duties were delegated away from SIDA. In the mid-1990s, a new structure emerged – a new joint Sida (Swedish International Development Co-operation Agency). Second, during the first two decades, SIDA operated on the basis of a number of themes, such as health, education, agriculture and industry. In addition, very intensive information activities took place, inspiring a lively debate on development issues. In the mid-1980s, the emphasis shifted from a thematical approach towards a more regional focus. With this, a discussion on roles and responsibilities between donors and recipients was intensified. At the same time, a downgrading of information resulted in aid becoming a less significant question to the Swedish public, even if a keen interest lived on within some NGOs and higher educational institutions.

It could also be added that while Sida is in charge of most of the *bilateral* co-operation, the Ministry of Foreign Affairs handles *multilateral* assistance. Lately, Swedish development co-operation has also been influenced by its membership of the European Union (EU) and the new emphasis on the former Soviet Union and Eastern Europe.

One controversial issue since the 1960s has been the selection of recipient countries. When SIDA was formed in 1965, six main recipients were selected, namely Ethiopia, Kenya, Tanzania, Tunisia, India and Pakistan. Of these, all but Tunisia and Pakistan are still major countries of co-operation. The process of abolishing Tunisia as a key country of co-operation points to many of the weaknesses in the mutual relationship, characterised by bad communication between the two parties – Sweden and Tunisia. No actual dialogue was instituted, and many directives from the Swedish side were unclear and open to misunderstanding. Obviously, this was not really a recipe for a closer future no-aid relationship (Wallberg 1982). One question is how much SIDA actually learned from this process, in an analysis of the existing unequal 'partnership'.

In Africa, the continued Swedish interest has shifted towards southern Africa, following the region's liberation from colonialism and white minority rule. Early on, this could be seen as an attempt to contribute to the establishment of a shield, built upon Botswana, Lesotho, Swaziland and Zambia, against southern Africa's apartheid regimes. Thereafter, adoption as a major country of co-operation has often followed from support for a national liberation movement, as happened in respect of Angola, Guinea-Bissau, Cape Verde, Mozambique, Zimbabwe, Namibia, and most recently the Republic of South Africa. During the 1990s, Uganda was added to the list: like so many other donors, Sweden became enchanted by President Museveni's populist appeal. Meanwhile, Botswana, Lesotho, Swaziland and Cape Verde have now been phased out.

The pattern in Asia is somewhat similar to that in Africa. With a starting point in the Indian subcontinent, Sri Lanka, India and Pakistan became early recipients. Following the formation of Bangladesh from East Pakistan, that new country was immediately included as a main recipient, while interest in Pakistan dwindled. Thereafter, the emphasis shifted towards countries emerging from long periods of war, such as Vietnam, Laos and to some extent also Cambodia. Afghanistan and the West Bank (Palestine) have now been added to this list.

Latin America has been somewhat more complicated in its relationship with Sweden. Two early countries of co-operation, Cuba and Chile, have lost this status. Aid to Chile was stopped very abruptly at the time of the Pinochet coup in 1973, while Cuba was always one of the most controversial countries of co-operation. The same applies in respect of Nicaragua. At present, interest in other countries in Central America and also Bolivia seems to be growing.

Prior to the 1990s, Portugal had been the only European country counted among main aid recipients. With the changes in Eastern Europe, this group has been expanded considerably to include the former Yugoslavia, Poland and the Baltic states among others.

Forty countries have been the principal recipients of Swedish aid, but the total number obtaining some form of aid is much larger. In 1996, no fewer than 114 countries had some form of bilateral aid relationship with Sweden. Tanzania, Mozambique, India, Nicaragua and Vietnam together received close to 30 per cent of the total. These countries have long been among the main recipients, even if this status has been surrounded by controversy. In most cases, the voices against the choice of countries for aid have argued that it leaned too far to the left of the political spectrum. According to Englund (1991), Sweden has chosen to ally itself with the enemies of our own enemy, e.g. the USA, which we are not able to attack directly. Therefore, we have found ourselves on the side of countries such as Cuba, Vietnam, Ethiopia and Angola. Englund does not mention Tanzania, but it has also been a controversial recipient. Numerous contributors to an anthology have elaborated on this issue (Rydén 1984). In many cases, these 'socialist' countries are said to have a poor record of development performance (but *cf.* Simon 1995).

Strong opposition in relation to other countries, e.g. Kenya, has been caused by a bad human rights record or the inability to achieve balanced regional development. On the one hand, we can question whether it is meaningful to give aid to Kenya, but on the other, in what way is it any worse than many other African nations? This brings us to the whole issue of power relationships between donors and recipients, and what kind of conditionalities can actually be imposed on aid delivery. The position on conditionalities is slowly being moved forward, parallel to calls for the recipients of aid to have a greater say.

Kalderén (1971) doubted market forces in stressing the need to develop the non-economic social sectors. Aid has to be a form of co-operation corresponding to demands from the recipients. Wohlgemut (1976) saw the establishment of SIDA as an opportunity for partnership to develop with people in the South, especially at the project level. At the same time, SIDA retained the right to assess projects before and after implementation, leaving it with strong central control. Around 1970, the concept of partnership emerged; this might actually not differ much from what is now re-emerging in the late 1990s. A measure of equality was to come from the method of country programming, which would give the Third World country an opportunity for long-term planning. However, at the same time, weak administrative systems made it difficult to implement the concept fully (Odén 1976).

Simultaneously with the debate on increased responsibility to the South, matters of conditionalities – strategic and commercial – gained importance. Edgren and Odén (1972) claimed that major donors were using this method as part of their global role in the economic and political power structure. It was suggested that SIDA should use the same leverage to achieve positive effects for Sweden, such as making it into a wholesale store for Swedish expert consultancies. Suddenly, we reached the situation of imposing commercial conditionalities. Rudengren (1976) made the point that if conflicts are to be avoided aid must be based on a commonality of values. With the donor as the stronger party, this might mean in reality that recipients adopt a Swedish value system – including an econometric conception of development. This is also coupled with the principle of increased commercialisation of aid (Dahlgren 1976). Officially, the connection between aid and private commercial interests was recognised in a government investigation into industrial assistance (Statens Offentliga Utredningar 1972). Gårdlund (1968) tried to put up a spirited defence of private investment in the Third World as being a positive development tool.

Many contributors to Rydén (1984) claimed that conventional aid is not at all efficient. They argued strongly for the private sector to be assigned a more substantial role. In their view, entrepreneurship and industrial competence were the factors that could put development assistance back on track. From this perspective, there should be more and not fewer conditionalities on aid. Blomberg (1991) has given numerous hints regarding what he regards as a profitable market – the aid business. If aid benefits us in the North, it will also function better in the South. This point directly contradicts the schools of dependency and structuralism. This mutual benefit thesis has recently been expressed again by Sida (1997: 17), claiming positive effects 'in both the short and long term to growth in Sweden':

> Development co-operation uses Swedish goods and services to a large extent. Some 60 per cent of Swedish support goes ultimately to Swedish companies and organisations. Through development co-operation Swedish companies can obtain an early foothold in new markets and can develop their know-how with new experience. It has been estimated that, in 1996, Sida's operations created 5,000 job opportunities in Sweden.

This quotation can be seen as one expression of the way that Swedish aid is shifting further from its own previous model. Another increasingly important component in Swedish aid is the NGOs (non-governmental organisations), for long something of a sacred cow in the development debate. At the same time as SIDA itself, a Swedish Peace Corps was established to cater for young professionals; it was replaced in the early 1980s by another structure, SVS (Swedish Volunteer Services – now called Forum Syd). Gyllensvärd and Sandberg (1989) introduced a somewhat self-critical approach to the NGO sector. The contributors claimed that a false and romanticised picture of development co-operation is only detrimental to solidarity itself. Therefore, all sectors of aid, including NGOs, have to be carefully scrutinised.

Trends in the NGO sector seem to be moving towards closer co-ordination with the rest of Swedish aid. In a government analysis (Statens Offentliga Utredningar 1994: 8–9), the directive was given that NGOs were also to account properly for their state funding. As official aid is being directed increasingly towards the basic economic preconditions for development, NGOs are partly being given the task of cushioning the immediate negative social effects of aid. In that respect, they have moved a long way from their original role of promoting alternative development thinking. As mentioned in Chapter 1, a possible missing link in development co-operation today is the active solidarity built up within the mainstream NGO sector during the late 1960s and early 1970s in protest against US aggression in Vietnam, the military atrocities in Chile, and the apartheid systems in Southern Rhodesia, Namibia and South Africa (*cf.* also the discussion on local power in Chapter 11 this volume).

A broad-ranging report by Riddell *et al.* (1995) made some overall assessments of Swedish NGOs. The findings are contrary to the popular view of NGO performance. On the positive side, we find that they are in most cases able to fulfil set objectives. However, if they are evaluated in relation to a more comprehensive development agenda, the picture becomes much gloomier. In relation to popular catchwords of the day, Swedish NGOs seem to be rather inefficient. Even if they tend to believe that they are working with the poor, it seems that they do not reach the poorest and probably not even the very poor. Similarly, an assumption is made that their projects are gender-sensitive. However, the report concluded that they might address the symptoms, but not the causes of poverty as these are linked to gender. Finally, the report expresses disappointment at the low level of local participation. All this would be a clear indication of the fact that NGOs have survived on an image of what is not there. They have not developed into an alternative that could harness popular solidarity; rather, they have become a second-class and inexpensive way to deliver development aid. Besides, given their increasing aid-based links to the state, to what extent are NGOs nowadays really 'non-governmental'?

Nyerere (1976) argued that the whole concept of aid is basically unhelpful, as development is mainly a task for people in the poor countries themselves. However, in a world where some are rich and others poor, it is an obligation and not charity to transfer resources from the rich to the poor. It was the rules of the international division of labour that had to be changed from being a relationship of exploitation to one of co-operation. Then it would be up to the Third World people to decide how to use their own resources.

New trends in development aid – towards an alternative agenda?

Neoliberalism or alternative thinking?

One dominant mode of thinking today is neoliberalism, which has been connected to the new conservatism of Thatcher and Reagan. In practical terms, it has to a large extent formed the backbone of policies formulated by the World Bank and the IMF (Hettne 1995: 112–114). To neoliberal thinkers, the main idea is to strengthen the role of the market by eliminating imperfections and interventions (Preston 1996: 254). A typical expression of this is when the World Bank (1990: 87) proposes to introduce certain measures for poverty alleviation, on condition that they do not disturb the macro-economic discipline. In a recent World Development Report (World Bank 1997a), the main issue was the role of the state, which should be reduced to a bare minimum while also exercising good governance.

Present world structures are a fertile breeding ground for the neoliberal development policies practised in most countries. A central feature of this is the debt crisis afflicting almost the entire Third World. At the same time, the collapse of the ideological system in the former Soviet Union and Eastern Europe created a new euphoria in extreme forms of capitalism and free-market economy. In the South, aid dependency and vulnerable regimes were left to follow dictates from outside, without considering the consequences. International trade was built up, benefiting the already affluent. The activities of multilateral and transnational companies were made easier, rather than being restricted by various agreements. In many cases, they damaged the environment beyond repair through overexploitation of natural resources.

Various development alternatives to this open capitalism have been suggested (Brohman 1996; Burkey 1993; Clark 1991; Ekins 1992). Often the emphasis is placed on the need for grassroots development and a bottom-up approach. Nabudere (1997) explores how the poor, who will search for their own post-traditional solutions, will resist the enforced development of modernisation. A central role in grassroots development has been given to the NGO sector, but numerous critical voices are being raised against that approach (Edwards and Hulme 1995; Fowler 1997; Tandon 1991; *cf.* also the comments on Swedish NGOs above).

It is on these two conflicting approaches that the development thinking of today is based. Below we will study how the potential alternatives are reflected in aid practice. As this is a comment on ongoing processes, only some brief hints regarding new trends will be given.

The DAC vision of the future

The DAC has introduced its own vision of the future in this period of general development pessimism. It still regards aid as a moral issue, while a reduction of prevailing underdevelopment is necessary for world security. In retrospect, it is claimed that internal initiatives in the South have been the main ingredient in the creation of development, but aid has been an essential complementary factor. Consequently, the DAC claims that aid works, giving examples such as the Green Revolution, decreasing birth rates, improved infrastructure, the reduced incidence of diseases and dramatically reduced poverty. Some countries are said to have prospered and achieved industrialisation, now standing on their own. On the other hand, countries with civil disturbances and bad governance have retarded development for a long time to come (OECD 1997: 12; see also Chapter 2 this volume).

In spite of the global trend of declining aid ratios, the DAC declares its commitment to contribute towards human and economic development. It seems a bit hollow in a world so embroiled in confrontation of various kinds to read the DAC proposition of 'a global development partnership effort through which we can achieve together the following ambitious but realistic goals (*ibid.*: 13). After almost half a century of development aid, it is interesting to note this continuous confidence expressed by the DAC, which is now focusing on the new key concept of partnership, basically to continue with a process of modernisation. Somehow, there is no indication that the DAC has adopted any of the debate on the meaning of development and alternative perspectives.

The reasonable goals that have been set up are:

- Economic well-being.
- Social development.
- Environmental sustainability and regeneration.

Among the sub-objectives to achieve this are universal primary education and access to appropriate health care by the year 2015. Somehow, these kinds of resolution do have an air of *déjà vu* around them. Now the vision includes a joint commitment between donors and recipients, improved co-ordination in respect of locally owned development strategies, and a determined effort to achieve better coherence between aid and other policies with an impact on developing countries. Whether this is a new policy or only a well-intentioned but watered-down vision, only the future can tell. Indeed, this seems to be a good example of the kind of international document discussed by Wolfe (1996) (see Chapter 1 this volume), in terms of which major concessions are necessary in order to yield a compromise so that it is possible to agree on something at all. As a further example of the recent debate, some notes from the Swedish scene will be sketched.

Extracts from a recent Swedish debate

As the end of the twentieth century approaches, internal debate on development co-operation has intensified. Cedergren (1997: 205) regards the main task of development aid as being to make itself superfluous. Consequently, the continued existence of aid is in itself a failure. According to the permanent secretary in the ministry of foreign affairs, Mats Karlsson, it is imperative that if aid is to lead to self-reliance, we have to start thinking about a no-aid relationship with Africa. Initially, this will lead to increased and not decreased volumes of aid, but in a whole new political framework (Karlsson 1996: 8–9). Within that kind of thinking, it is also necessary for African regimes to accept their responsibility, based on the involvement of civil society in a democratic process of change, something that runs counter to the 'Washington consensus' vision of democracy.

A similar point was made by the then director-general of the Norwegian Agency for Development Co-operation (NORAD), Per Grimstad, in stressing the responsibility of the recipients. At present, Norwegian aid could contribute to the co-ordination and allocation of funding, but in the long run the aim must be to make a withdrawal possible (Grimstad 1996: 22–23).

During the last few years, SIDA and other smaller bodies have been transformed into an integrated Sida through a major organisational restructuring. At the same time, it has been suggested that the principal activities be focused on themes, as elaborated on in four action plans, dealing with poverty alleviation, gender, sustainability and democracy/ human rights respectively. Once again, it is difficult to see how the new policy directions are in any way related to comprehensive development research. Furthermore, the four plans might express a number of desirable aims on their own, but no attempt is being made to integrate them. Contrary to the required professionalism, the new policies appeal to idealism and morally correct humanitarianism. The need to position aid within a more integrated understanding of what development is, as discussed above, seems to be being totally ignored.

A central question is whether the topics chosen are a response to popular Swedish opinion, or a local agenda in the Third World? According to Tandon (1991: 75), NGOs have pushed three specific issues with a clear European–American value bias in Africa – namely gender, human rights and ecology. It seems to be more than coincidental that Tandon mentions what are now three out of the four main themes for Sida's future work. Similarly, Nyerere (1998: 26) reacted strongly to the aid-related use of 'good governance', calling it 'arrogant and patronizing ... like a tool for Neo-colonialism'. This is not to say that these sectors have been formulated with devious intentions. However, if donors were to be more attentive to various voices from the South, some other key issues might have been selected.

Another concept currently at the heart of the Swedish debate on the Third World is partnership. This stems from the idea that recipients will themselves be responsible for their projects. Partnership has been introduced for Africa, and will soon be a topic of discussion in relation to Asia. According to this approach, there will not be any donors or recipients, only partners in development – comprising governments as well as the commercial sector, civil society, NGOs and voluntary organisations. It will mean that a new spirit of understanding and respect has to be created (Ministry of Foreign Affairs 1997) and that the aid agencies are able to enter a phase of competence building in dialogue not only with African representatives but also within Swedish society. After all, development not only matters locally in the Third World but also represents a dilemma facing global relationships. Here we find one of the few examples in which the conflictual aspects of development are actually addressed at the policy level. The intention of this kind of thinking does seem promising, but the difficult task is implementing it (Kayizzi-Mugerwa, 1998). One striking omission from a policy that is supposed to cover all relations between Sweden and Africa is the export of arms.

In a paper relating to the 1999 budget, some aspects of the practical application of development co-operation are outlined, e.g. the building of knowledge and capacity. This is supposed to be related to capacity in the South, among ourselves in Sweden and internally at Sida (Sida 1998). However, even if there is a strong ambition to learn, there is still a touch of Eurocentrism in this thinking. The crucial issue of how we are going to learn from the South, and how one part of the South is to learn from another, is not really addressed. Furthermore, co-operation with Swedish universities, civil society and not least the immigrant communities must be emphasised much more if we are to build competence in Sweden that can launch a broader dialogue with the South. Swedish society must also reconsider its roots and background (see Chapter 1) to be able to contribute fully to the international debate.

Conclusions

Aid is a controversial topic, and the contrasting views in the debate tend to be painted rather black or white. Sometimes actions taken by donors are more a response to negative critics than a response to a development crisis. Basically, aid is supposed to be built on a humanitarian attempt to contribute towards a positive modernisation process in the South. How to make this as efficient as possible is a key issue. Initially, the view was that traditional cultures and values represented obstacles to development. Today, many might still think the same thing, but it is not politically correct to use such terminology.

In terms of aid volumes, this has turned into a virtual World Cup race between various donors to show their national commitments. The leading position taken by the USA has long since been replaced by the Scandinavian countries and the Netherlands. Often a positive feeling towards aid has been created by a particular strong political personality in the North. The basic humanitarian motive for aid has been challenged from various angles. It is claimed that aid is actually more a response to strategic, political and commercial interests, which are also reflected in the selection of recipients.

A prevailing critique of aid is that there seems to be no comprehensive understanding of what development really is. In many cases, projects are assessed in a positive way in their own right, but they do not seem to contribute towards a broader development process. It has been claimed that there is no actual evidence to show that aid leads to any positive development. Indeed, evaluation results are rather contradictory. Coupled with the fact that concealed interests are involved in the formulation of development policies, this calls for new forms of dialogue between researchers and practitioners, among others. In this debate, the normal social science objective of 'scoring points' has to be replaced by a desire to build new knowledge. Even the present dominance of econometric solutions to all development dilemmas might be an effect of the preoccupation among alternative researchers to criticise aid rather than attemping to find alternatives.

So far, the influence of current development thinking on aid has been quite fragmentary. While large parts of the modernisation paradigm and associated growth theories have been easy to adopt, the bulk of structuralist research has not provided a basis for development practice. This can be seen largely as a consequence of the fact that the conflictual way of thinking is difficult to adjust to as the basis for co-operation. It is much more convenient to assume that aid can be built on harmony. Unfortunately, the real world is not so easy to deal with.

As development research is not providing any concrete, simple path to development, it opens up the possibility for donors to select what they feel at ease with. However, the will to do good is soon overtaken by the need for a deeper analysis of the complex development environment of which aid forms a part. This leads to a situation where it is often claimed that a new agenda is being followed, but only some cosmetic changes have in reality been made. In the same way, a research vocabulary might be adhered to, but the meaning of various concepts is open to individual (re)interpretation.

Furthermore, the knowledge base and actual understanding of what is happening on the ground is limited in various ways. A refined academic discussion often alienates the research community from local realities. Short-term consultancies often replace previous broad-based development research. Indigenous researchers are often lured by substantial financial inducements to say 'the right thing' if they are to get any more consultancy assignments.

It seems that new trends in development aid during the 1990s are not built on previous experiences to any large degree. After fifty years of development aid, the underlying concepts and methods have not changed fundamentally, even if the donor organisations would strongly disagree with this statement. However, donors and recipients are still entangled in the same kind of structural relationship, even if the vocabulary has changed to include words such as 'empowerment', 'partnership' and 'participation'.

Expressing these concerns is not meant to perpetuate confrontation but to open the door to new forms of dialogue and reflection. Understanding of the processes of development, as well as the future (no-)aid situation, urgently requires this.

References

Adams, N. (1993) *Worlds Apart – The North–South Divide and the International System.* Zed Books, London and New Jersey.

African Business (1988) May, London.

Amland, B. (1993) *Bistånd eller Børs? Næringlivets roll i norsk u-hjelp.* J.W. Cappelens Forlag, Lillehammer.

Andersson, A., Heikensten, L. and **de Wylder, S.** (1984) *Bistånd i Kris – en bok om Svensk u-landspolitik.* LiberFörlag, Stockholm.

Andersson, L. (1989) *Biståndet till u-länderna.* SIDA:s Kursgård, Uppsala.

Bauer, P. (1984) *Biståndsmyten.* Timbro, Stockholm.

Bendix, P. (1996) Exemplary in concept and reach – the development policy of the Netherlands, *Development and Co-operation* 3: 24–26.

Bigsten, A. (1984) Vad är effektivt bistånd? In Andersson, C., Heikensten, L. and de Vylder, S. (eds) *Bistånd i Kris – en bok om svensk u-landspolitik*, LiberFörlag, Stockholm, 61–75.

Blomberg, P. (1991) *Biståndsmarknaden – affärsmöjligheter inom svenskt och internationellt bistånd*. Sveriges exportråd, Stockholm.

Booth, D. (1993) Development research: from impasse to a new agenda. In Schuurman, F. (ed.) *Beyond the Impasse – New Directions in Development Theory*, Zed Books, London, 49–76.

Booth, D. (ed.) (1994) *Rethinking Social Development – Theory, Research & Practice*. Longman, Harlow.

Brandt, W. (1980) *North–South – A Programme for Survival*. Pan, London.

Brandt, W. (1983) *Common Crisis – North–South: Co-operation for World Recovery*. Pan World Affairs, London and Sydney.

Brohman, J. (1996) *Popular Development – Rethinking the Theory and Practice of Development*. Blackwell, London and Oxford.

Burkey, S. (1993) *People First – A Guide to Self-Reliant, Participatory Rural Development*. Zed Books, London and New Jersey.

Cassen, R. and Associates (1986) *Does Aid Work?* Clarendon Press, Oxford.

Cedergren, J. (1997) Framtidsperspektiv. In Wohlgemut, L. (ed.) *Bistånd på Utvecklingens Villkor (Ny reviderad Upplaga)*, Nordiska afrikainstitutet, Uppsala, 201–206.

Chambers, R. (1997) *Whose Reality Counts – Putting the First Last*. Longman, Harlow.

Clark, J. (1991) *Democratizising Development – The Role of Voluntary Organizations*. Earthscan, London.

Cooper, C. and **van Themaat, J.V.** (1989) Dutch aid determinants, 1973–85: continuity and change. In Stokke, O. (ed.) *Western Middle Powers and Global Poverty: The Determinants of the Aid Policies of Canada, Denmark, the Netherlands, Norway and Sweden*, The Scandinavian Institute for African Studies, Uppsala, 117–156.

Cumming, G. (1996) British aid to Africa: a changing agenda? *Third World Quarterly* **17**(3): 487–501.

Curtis, M. (1997) United Kingdom. In Randel, J. and German, T. (eds) *The Reality of Aid – An Independent Review of Development Co-operation 1997–1998*, Earthscan, London, 144–152.

Dahlgren, G. (1976) U-landspolitik och biståndet. In Wohlgemut, L. (ed.) *Bistånd på mottagarens villkor – filosofi och teknik*, SIDA, Stockholm, 7–15.

Dube, S. (1988) *Modernization and Development – The Search for Alternative Paradigms*. The UN University, Tokyo, and Zed Books, London and New York.

Dumont, R. (1962) *L'Afrique noire est mal parti*, Editions du Seuil, Paris.

Edgren, G. (1984) Om biståndets villkor. In Andersson, C., Heikensten, L. and de Vylder, S. (eds) *Bistånd i Kris – en bok om svensk u-landspolitik*. LiberFörlag, Stockholm, 39–53.

Edgren, G. and **Odén, B.** (1972) *Biståndet som hävstång*. Prisma, Stockholm.

Edwards, M. and **Hulme, D.** (eds) (1995) *Non-Governmental Organisations – Performance and Accountability: Beyond the Magic Bullet*. Earthscan, London.

Ehrenpreis, D. (1997) Fungerar biståndet? – en genomgång av 30 års erfarnheter. In Wohlgemut, L. (ed.) *Bistånd på Utvecklingens Villkor (Ny reviderad Upplaga)*, Nordiska afrikainstitutet, Uppsala, 14–30.

Ekins, P. (1992) *A New World Order – Grassroots Movements for Global Change*. Routledge, London and New York.

Englund, R. (1991) *Till vänster om marknaden – bistånd med slagsida*. Timbro, Stockholm.

Eriksen, T.H. (ed.) (1989) *Hvor mange hvite elefanter?* ad Notam, Oslo.

Eriksen, T.L. (ed.) (1987) *Den Vanskelige Bistanden – Noen Trekk ved Norsk Utviklingshjelps historie*. Universitetforlaget, Drammen.

Fowler, A. (1997) *Striking a Balance – A Guide to Enhancing the Effectiveness of Non-Governmental Organisations in International Development*. Earthscan, London.

Frank, A. (1991) No escape from the laws of world economics, *Review of African Political Economy* **50**: 21–32.

Garbo, G. (1993) *Makt og bistånd – En ambassadors mote med norsk bistandspolitik i Afrika*. Spartacus Forlag, Oslo.

Garbo, G. (1997) Norway. In Randel, J. and German T. (eds) *The Reality of Aid – An Independent Review of Development Co-operation 1997–1998*. Earthscan, London, 113–118.

Gårdlund, T. (1968) *Främmande investeringar i u-land*. Almquist & Wiksell, Stockholm.

Grimstad, P. (1996) Recipient responsibility – developing countries must chart their own development, *Development and Co-operation*, **2**: 22–23.

Gyllensvärd, E. and **Sandberg, S.** (eds) (1989) *Folkets Bistånd? En debattantologi om Folkrörelsernas U-landsarbete*. Svensk Volontärsamverkan, Stockholm.

Hadjor, K. (1993) *Dictionary of Third World Terms*. Penguin, Harmondsworth.

Hancock, G. (1991) *Lords of Poverty*. Mandarin, London.

Hayter, T. and **Watson, C.** (1985) *Aid – Rhetoric and Reality.* Pluto Press, London and Sydney.

Hettne, B. (1995) *Development Theory and the Three Worlds* (2nd edn). Longman, Harlow.

Jacoby, E. (1971) Den gröna revolutionen och det svenska biståndet. In *U-debatt-Om mål och metoder i biståndsarbetet,* Rabén & Sjögren, Stockholm, 192–198.

Jensen, J. and **Nielsen, M.** (1981) *Hvad er COMECON?* Sydjysk Universitetforlag, Esbjerg.

Kaldéren, L. (1971) Inledning. In *U-debatt-Om mål och metoder i biståndsarbetet,* Rabén & Sjögren, Stockholm, 7–30.

Karlsson, M. (1996) Preface. In Havnevik, K. and van Arkadie, B. (eds) *Domination of Dialogue? – Experiences and Prospects for African Development Co-operation,* Nordiska afrikainstitutet, Uppsala, 7–12.

Karlström, B. (1996) *Det omöjliga biståndet. Andra Upplagan.* SNS Förlag, Stockholm.

Kayizzi-Mugerwa, S., Olukoshi, A. and **Wohlgemut, L.** (1998) *Towards a New Partnership with Africa – Challenges and Opportunities.* Nordiska Afrikainstitutet, Uppsala.

Kihlberg, M. (1971) Forskning och modernisering. In *U-debatt-Om mål och metoder i biståndsarbetet,* Rabén & Sjögren, Stockholm, 301–315.

Knutsson, K. (1965) *Tekniskt bistånd i traditionella samhällen.* Nordiska afrikainstitutet, Uppsala.

Kungl. Maj:t. (1962) *Proposition nr 100 Dr 1962.*

Kungl. Maj:t. (1968) *Proposition nr 101 Dr 1968.*

Long, N. and **Long, A.** (eds) (1992) *Battlefields of Knowledge – The Interlocking of Theory and Practice in Social Research and Development.* Routledge, London.

Markensten, K. (1967) *Svensk u-landshjälp idag.* Almquist & Wiksell, Stockholm.

Ministry of Foreign Affairs (1997) *Partnership with Africa: Proposals for a New Swedish Policy towards Sub-Saharan Africa.* Stockholm.

Morton, J. (1996) *The Poverty of Nations – The Aid Dilemma at the Heart of Africa.* J.B. Tauris Publishers, London and New York.

Nabudere, D. (1997) Beyond modernization and development, or why the poor reject development, *Geografiska Annaler, Series B, Human Geography,* **79B**(4): 203–215.

Nyerere, J. (1976) Mottagarens syn på biståndet. In Wohlgemut, L. (ed.) *Bistånd på mottagarens villkor – filosofi och teknik,* SIDA, Stockholm, 29–43.

Nyerere, J. (1998) Governance in Africa, *Southern African Political and Economic Monthly* **11**(6): 26–28.

Odén, B. (1976) Landprogrammering – biståndsteknik och biståndsfilosofi. In Wohlgemut, L. (ed.) *Bistånd på mottagarens villkor – filosofi och teknik.* SIDA, Stockholm, 45–55.

Odén, B. (1984) Svenskt bistånd rättar in sig i OECD-ledet. In Andersson, C., Heikensten, L. and de Vylder, S. (eds) *Bistånd i Kris – en bok om svensk u-landspolitik.* LiberFörlag, Stockholm, 17–26.

OECD (1997) *Efforts and Policies of the Members of the Development Assistance Committee Development Co-operation.* Paris, DAC 1996 Report.

Patterson, R. (1997) *Foreign Aid after the Cold War – The Dynamics of Multipolar Economic Competition.* Africa World Press, Trenton and Asmara.

Pearson, L. (1969) *Partners in Development – Report of the Commission on International Development.* Praeger, New York.

Porter, D., Allen, B. and **Thompson, G.** (1991) *Development in Practice – Paved with Good Intentions.* Routledge, London and New York.

Preston, P.W. (1996) *Development Theory – An Introduction.* Blackwell, Oxford.

Raffer, K. and **Singer, H.** (1996) *The Foreign Aid Business – Economic Assistance and Development Co-operation.* Edward Elgar, Cheltenham and Brookfield.

Randel, J. and **German, T.** (eds) (1996) *The Reality of Aid 1996 – An Independent Review of International Aid.* Earthscan, London.

Randel, J. and **German, T.** (eds) (1997) *The Reality of Aid 1997/8 – An Independent Review of International Aid.* Earthscan, London.

Riddell, R. (1987) *Foreign Aid Reconsidered.* James Currey and ODI, London.

Riddell, R., Bebbington, A. and **Peck, L.** (1995) *Promoting Development by Proxy – An Evaluation of the Development Impact of Government Support to Swedish NGOs.* SIDA Evaluation Report, Stockholm.

Riksrevisionsverket (1988) *Lär sig SIDA? En granskning av SIDAs förmåga att lära sig av erfarenheterna.* Stockholm.

Rist, G. (1997) *The History of Development – From Western Origin to Global Faith.* Zed Books, London and New York.

Rostow, W. (1960) *The Stages of Growth – A Non-Communist Manifesto.* Cambridge University Press, Cambridge.

Rudengren, J. (1976) Svensk biståndspolitik. In Wohlgemut, L. (ed.) *Bistånd på mottagens villkor – filosofi and teknik,* SIDA, Stockholm, 17–27.

Rugumamu, S. (1997) *Lethal Aid – The Illusion of Socialism and Self-Reliance in Tanzania.* Africa World Press, Trenton and Asmara.

Rydén, B. (1984) *Bistånd under omprövning.* SNS Förlag, Stockholm.

Sachs, W. (1992) *The Development Dictionary,* Zed Books, London.

Schleicher, H (1995) Promoting Socialism and Anti-Colonialism B. The Development Policy of the GDR, *Development and Co-operation* **6**: 23–25.

Sida (1997) *Sida Looks Forward: Sida's programme for global development.* Stockholm.

Sida (1998) *Sidas Budgetunderlag 1999,* Stockholm.

Simon, D. (1995) The demise of 'socialist' state forms in Africa: an overview, *Journal of International Development* **7**(5): 707–739.

Soja, E. (1989) *Postmodern Geographies – The Reassertion of Space in Critical Social Theory.* Verso, London and New York.

South Commission (1990) *The Challenge to the South.* Oxford University Press, Oxford.

Statens Offentliga Utredningar (1972) *Industriutveckling och utvecklingssamarbete.* Betänkande avgivet av Industribiståndsutredningen, Allmänna Förlaget, Stockholm, 90.

Statens Offentliga Utredningar (1973) *Forskning för Utveckling.* Betänkande avgivet av U-landsforskningsutredningen, Stockholm, 41.

Statens Offentliga Utredningar (1994) *Analys och utvärdering av bistånd.* Betänkande av kommittén för analys av utvecklingssamarbete, Stockholm, 102.

Stokke, O. (ed.) (1989) *Western Middle Powers and Global Poverty – The Determinants of the Aid Policies of Canada, Denmark, the Netherlands, Norway and Sweden.* The Scandinavian Institute of African Studies, Uppsala.

Tandon, Y. (1991) 'Foreign NGOs, uses and abuses: an African perspective, *IFDA Dossier,* April/June, Geneva: 67–78.

U-debatt-Om mål och metoder i biståndsarbete (1971) Rabén & Sjögren, Stockholm.

Utenriksdepartementet (1992) *Om utviklingstrekk i Nord–Sør forholdet og Norges samarbeid med utviklingslande,* St. meld, nr 51 (1991–92). Oslo.

Wallberg, B. (1982) *Sverige – Tunisien 1962–1982 – Erfarenheter av 20 års utvecklingssamarbet.* SIDA, Stockholm.

Wildeman, C. (1997) The Netherlands. In Randel, J. and German, T. (eds) *The Reality of Aid – An Independent Review of Development Co-operation 1997–1998.* Earthscan, London, 98–104.

Wohlgemut, L. (ed.) (1976) *Bistånd på mottagarens villkor – filosofi och teknik.* SIDA, Stockholm.

Wolfe, M. (1996) *Elusive Development*, Zed Books, London.

World Bank (1986) *World Development Report 1986*. Oxford University Press, Washington.

World Bank (1990) *Making Adjustment Work for the Poor – A Framework for Policy Reform in Africa*. The World Bank, Washington DC.

World Bank (1997a) *World Development Report*. Oxford University Press, Washington.

World Bank (1997b) *World Development Indicators 1997*. The World Bank, Washington DC.

World Bank (1998) *World Development Indicators 1998*. The World Bank, Washington DC.

World Commission on Environment and Development (1987) *Our Common Future*. Oxford University Press, Oxford.

Yu, G. (1975), *China's African Policy – A Study of Tanzania*. Praeger, New York.

CHAPTER 8

Continuity and change in the Netherlands–Mali bilateral aid relationship

A.A. de Jong, A.C.M. van Westen and E.J.A. Harts-Broekhuis
University of Utrecht

Introduction

Dutch development co-operation with countries of the Third World has undergone considerable change during the last few decades. This is true with respect to its objectives, the size of the aid budget and its composition, as well as its geographical distribution. Thinking about development aid has also drastically changed during this period among politicians, scholars and the general public in the Netherlands. This contribution focuses on the relationship between this 'thinking' and the practice of development aid through a case study of bilateral development co-operation between the Netherlands and Mali during the period 1975–1996. Key questions are:

- To what extent have changes in the Netherlands–Mali aid relationship been affected by the evolution of development theory?
- How have domestic policy concerns and public opinion guided the aid relationship?
- What has been the impact of changing trends in multilateral development policy advanced by institutions such as the World Bank, the International Monetary Fund (IMF), the European Commission and the Comité Interétats de la Lutte contre la Sécheresse dans le Sahel (CILSS)?
- To what extent has the Malian side been able to influence the aid relationship?
- What are the achievements and shortcomings of Dutch aid to Mali?

Although the nature of Dutch aid contacts varies enormously among recipient countries, the Malian case may be used to gain an understanding of the factors behind changing aid practice. In the first section of this chapter, development assistance is placed in the context of changing development thinking. The second section gives a brief review of Dutch development co-operation, demonstrating that trends were often derived from other donors. In the third section, changes in the development relationship with Mali are discussed, with a view to seeing how formal principles match reality; practice was often more sustainable than termi-nology. In the last section, some major points of discussion with respect to the leading questions will be presented.

Theoretical viewpoints on development and aid

In mainstream development thinking, different paradigms have alter-nated and sometimes existed side by side, some of them more influential on the practice of development assistance than others.

The historical background of development thinking in the Netherlands cannot be separated from the colonial involvement in Indonesia and the West Indies, i.e. Suriname and the Netherlands Antilles. Although the independence of Indonesia in 1949 changed the spatial focus of develop-ment assistance, the ideas behind the efforts were not fundamentally af-fected. The (former) colonies were seen as 'lagging behind' in development, a backwardness that could be overcome by capital invest-ment and the spread of knowledge. Aid could contribute in both respects. These ideas were based on the linear and evolutionary development para-digm (or 'modernisation' thinking), influential in the 1950s and 1960s. The motive for development assistance in these years was certainly not only altruistic. This was reflected when in the 1960s, the accent in Dutch aid shifted from multilateral to bilateral aid. Organisations of employers insisted on supplying more aid through bilateral channels in order to stimulate exports, and to increase the purchasing power of undeveloped markets in the long term, and also to secure access to natural resources (Lieten and van der Velden 1997).

At the end of the 1960s, ideas on the role of development aid became more diverse, when the heterogeneity of the 'less developed' world was better understood. Increasingly divergent development experiences between countries and within countries drew attention to the processes of polarisation and of marginalisation. The ensuing structuralist paradigm of the 1970s, which focused on 'dependency' and 'imperialism', did not provide an easy framework for development co-operation. The pre-occupation with inequality and exploitation by countries in the world core of those in the periphery sometimes challenged official development co-operation as a ploy to preserve neocolonialist relations. On the other hand, the structuralist view did contribute by encouraging interest in inequality and poverty. This was reflected in the 'basic needs' approach aiming at combating poverty. It was also instrumental in boosting the role of idealistic and critical non-governmental organisations in the aid arena (Schrijvers 1997). Thus, the number of development agencies (NGOs) and the development budget increased rapidly during this period.

At the end of the 1970s, new topics from different backgrounds were added to the development agenda. The gender bias of ongoing activities was criticised, followed by increasing awareness of ecological issues. As a result, the objectives of development efforts were broadened, and the concept of sustainable development was introduced.

The 1980s saw the maturing of development co-operation. Changes in approach stemmed from a synthesis of earlier paradigms rather than from an entirely new vision. Pragmatism took over from idealism, in many cases imposed by worsening economic conditions, which led to the structural adjustment programmes (SAPs) imposed by the IMF and World Bank. Although not a new concept of development, the stark 'new realism' of the 1980s entailed a considerable shift in approach in development policy. The underlying idea of 'adjustment' is that the market is the best instrument to guide the allocation of resources for economic development; the state should refrain from the proactive role typical of previous decades and try to play an 'enabling role' as regulator and intermediary. In the political field, decentralisation of public services and democratisation were widely pursued. As a consequence, conditions in many developing countries changed rapidly, and new forms of development assistance were needed. Structural adjustment required financial support for macro-economic reforms, and assistance for local initiatives and social support organisations became more important.

The 1990s show no break with the preceding decade in terms of development philosophy, but nevertheless they have seen some important modifications. A major change took place in the political context of aid with the collapse of the Soviet bloc and the ensuing claimed 'end of ideology'. Political objectives, previously a major consideration in aid, have lost much of their meaning. Economic considerations, on the other hand, have gained in importance in an increasingly globalised world economy, where private (commercial) capital flows have rapidly overtaken stagnant aid flows. The precise outcome of such transformations varies, but in general terms, weaker players in the global arena – poor peripheral countries, including Mali – tend to lose ground, while the more dynamic developing countries of Latin America and East Asia move ahead, though not without problems. While aid has become less important as a tool for development, it is also changing in composition. The preoccupation with macro-economic stability in the 1980s is complemented by a concern for 'human development', supported by annual UNDP (United Nations Development Programme) reports since 1990. In this vein, assistance again aims more specifically at poverty alleviation, and local popular participation and equality of opportunity are stressed. These fit well with explicit consideration of environmental and gender issues. NGOs continue to play an important role, and local authorities are taken more into consideration than before. These shifts in development policy in the 1990s are further encouraged by the ascendancy of postmodernism. Postmodernist ideas may present a watershed in development thinking, challenging authoritative knowledge, technical–rational arguments and the role of experts (Preston 1996; Simon Chapter 2, this volume). From this angle, development is viewed as a flexible, open-ended process with many possible directions, which are determined more by the actors involved – including the public sector – than was envisaged in the traditional functionalist or structuralist approaches (Schrijvers 1997). How postmodern ideas as such will affect the development practice – indeed, whether they will have a significant impact at all – is as yet difficult to assess.

A short review of the history of Dutch development co-operation

Objectives and channels

Over the nearly fifty years since its inception, Dutch foreign aid has evolved from a simple instrument to stimulate economic growth in its former colonies to a complex of activities in many countries, representing a substantial amount of money: the 'development industry' has matured. When Truman mentioned the support for developing democracies as 'Point Four' in his inaugural speech (in January 1949), and thus ushered in the era of modern development co-operation, the Netherlands was among the early nations to follow his example. According to Hoebink (1997), this was prompted in part by the welcome opportunity to make good use of the tropical expertise developed in the colonial territories of Indonesia and Suriname, for which there was little use at home. Policy makers and administrators were preoccupied in particular with stimulating economic growth by modernising the rural economies.

Decolonisation at first encouraged the use of multilateral channels. From the early 1960s on, following the example of other European nations and at the insistence of the Dutch private sector, efforts focused more on bilateral aid. In 1963, in the wake of Kennedy's Peace Corps, a volunteer service was created, SNV (Stichting Nederlandse Vrijwilligers or Dutch Volunteer Foundation), a popular organisation in times of development optimism and simple solutions. This organisation still exists under the same acronym, but it is now a part of channelling official Dutch aid, working with professionals, implementing relatively small-scale and 'soft' social programmes. Well into the 1970s, domestic strengths and sectoral priorities were a major force in shaping the Dutch co-operation effort: port development, land reclamation, construction of dams and river embankments, irrigation schemes, fertiliser supply, and veterinary services are the main examples. Sometimes the zeal of exporting national hallmarks had odd results, such as bicycle lanes in countries not yet aware of the need for them (Burkina Faso), herds of Frisian cows depending on air-conditioned stables for their survival (Mali), or cold storage for potatoes. But there have also been appropriate and successful initiatives. In recent decades, the focus on national specialisms has declined, thanks to the steady professionalisation of the aid industry. Persistently, however, rural development has stood at the centre of Dutch development efforts, with urban initiatives gaining cautious support only in the 1990s.

A major transformation in foreign aid since the early years has been the redefinition of target groups and themes. The World Bank's 'discovery' of poverty as a prime focus for intervention (in McNamara's 1973 Nairobi speech, calling for *redistribution with growth*) was soon taken up by the left-leaning Dutch government that assumed power in the same year. Jan Pronk, its energetic minister for development co-operation (1973–1977, and again since 1989), is credited with having equipped Dutch foreign aid with its first fairly comprehensive and explicit policy perspective. In his first term as minister, the old preoccupation with the technicalities of economic growth – trying to fill the 'savings gap' between domestic savings and the capital investment needs of developing countries – was broadened to encompass a concern with poverty and the distribution of income and access to other resources. The recipient society was no longer viewed as homogeneous, but its internal stratification in terms of power and income was recognised as a key development issue. The basis was laid for the poor as target group and for basic needs approaches (Cooper and van Themaat 1989). This has remained a major theme in the Netherlands' foreign aid programmes, at least on paper, but with regular shifts of emphasis by successive cabinets. Generally, these shifts continued to follow international trends – mostly World Bank views – as has often been observed among the smaller donor countries. Within the domestic political scene, relatively conservative ministers have tended to stress the role of the Dutch private sector more than the politicians of relatively progressive coalitions. Thus, a preoccupation with poverty eradication in the mid-1970s was succeeded by a two-pronged approach around 1980: poverty eradication *and* promotion of economic self-reliance. This in turn has given way to the single objective of structural poverty eradication since the mid-1980s, edging closer to the structural adjustment approach advocated by the World Bank as a response to the Third World debt crisis. However, when structural adjustment was the point of departure of much of the international aid scene at the height of the neoliberal era in the late 1980s, the Netherlands both embraced this 'Washington consensus' and tried to remain on the humanitarian side of it (Sterkenburg and van der Wiel 1995). In due course, the pendulum swung back somewhat, when even the World Bank recognised the negative social costs of strict adjustment policies. It is at this point that Dutch policy, together with like-minded countries, is credited with having had some influence on multilateral policies, rather than merely resigning itself to a position at the receiving end of policy innovations. As a result, poverty alleviation and environmental concerns – sustainability – have gained in stature on the multilateral agenda. The 'Human Development' approach of the UNDP since 1990 drawing attention to people's well-being following the exclusive focus on economic macro-indicators of the no-nonsense 1980s, is reflected in the central Dutch development goal of the 1990s, namely sustainable poverty eradication, incorporating development objectives such as gender equality, (ecological) sustainability, human rights and good governance. This policy perspective was elaborated in two government white papers: *A World of Difference* (Ministerie van Buitenlandse Zaken 1990) and *A World in Dispute* (Ministerie van Buitenlandse Zaken 1993). These publications prepared the way for a more profound reorientation in the mid-1990s.

In an attempt to take account of the impact of progressive globalisation, the transformation of the former socialist countries and deepening European integration within the EU, the Dutch government embarked on a general reorientation of its foreign policy in 1994. An important outcome of this process was the 'decompartmentalisation' of development cooperation and other aspects of foreign policy (Lensink 1995). This operation departed from the recognition that development policy is more and more embedded in other policy concerns in a progressively more integrated world in terms of economic, political and cultural linkages. Hence, the argument goes, 'traditional' development co-operation should not be kept artificially separated from other legitimate policy concerns, including social services provision, catering to refugees, environmental concerns, human rights, democratisation and good governance, peace keeping, and support for reform in the transitional economies: all are seen as issues for a broader international co-operation framework. In short, development policies are to be integrated with other foreign policies. At the end of 1995, the first phase of the 'regauging' of foreign and development co-operation policy was concluded. This included some important institutional changes, including the reorganisation of the ministry of foreign affairs according to a 'matrix' structure, with regional subdivisions on the one hand and subject matter subdivisions on the other. The Netherlands has a cabinet minister for 'development co-operation', but his staff and his budget are integrated with those of the ministry of foreign affairs. Of specific importance to development policy now is the closer integration of economic policy and foreign aid within the ministry, as well as a significant decentralisation of decision making on aid from the central bureaucracy in The Hague to the embassies. This necessitated the transfer of staff to the embassies, and the recruitment of local staff in the recipient countries to handle the aid programmes locally.

This localisation of the administration of the foreign aid programme follows an earlier delegation of development tasks in 1985, when embassies were assigned specialists for specific development sectors. Thus, Dutch embassies not only have diplomatic and consular tasks but are also responsible for the management and, since 1996, for the implementation of foreign aid. The increased localisation of management, administration and specialist consultancy work is seriously affecting the 'aid industry' at home, a source of considerable opposition (Zoomers 1997). On the other hand, increased local spending may be expected to have a positive impact on the recipient country's economy. The retention rate of Dutch development spending – i.e. the share flowing back to Dutch businesses or households – is already relatively low in comparison with most donors, at 45–50 per cent in the early 1990s, and probably lower now (Hoebink 1997). At the same time as the decentralisation programme, a shift away from projects as an aid modality in favour of programme assistance is being implemented, aiming at more consistent longer-term interventions at lower administrative costs. This is notwithstanding the mixed results that some multilateral agencies, e.g. UNDP and UNOPS (United Nations Office of Projects Services) have had with similar reforms. Overall, the consequences of the reorganisation for the Dutch aid budget cannot as yet be completely evaluated.

Aid budget and geographical focus

A quantitative target for Dutch foreign aid was first established in 1965, at 1 per cent of net national income, a target that was first met in the early

1970s. This respected the UN target of dedicating at least 0.7 per cent of GNP to foreign aid, which has consistently been met since then, an achievement shared only with the Scandinavian donor countries (Lensink 1995). In 1973, the target was raised to 1.5 per cent of net national income, and this was effectively realised as from 1976. In 1993 the target was reduced, resulting in lower expenditure levels (in relative terms) since then. Actual quantitative official development assistance (ODA) performance levels of Dutch foreign aid cannot readily be derived from these budget data, since criteria differ from those used by UN agencies and the OECD. For instance, administration costs and several non-ODA budget items (e.g. expenditures on refugees in the Netherlands, funds for encouraging trade with developing countries, part of the acquisition costs of transportation aircraft for the Dutch air force, which are used for humanitarian relief supplies, etc.) are included in the Dutch development budget.

In terms of its geographical distribution, Dutch aid efforts have increasingly been spread over a growing number of countries. This is clearly demonstrated in the declining share of total ODA allocated to the ten most important recipient countries in Table 8.1. The fall was exacerbated in the early 1990s due to Indonesia's ending of its co-operation relationship with the Netherlands. This steady spread has taken place in spite of several policy initiatives to concentrate on a smaller number of countries in the interests of building up a more meaningful 'development relationship'. It is accepted that a small donor country such as the Netherlands cannot effectively operate in all corners of the globe. At the same time, however, pressures to spread the budget ever more widely appear irresistible. Such pressures may come from the Dutch corporate sector, since the existence of an aid relationship of some sort often makes entry to various national markets easier, especially when it comes to the acquisition of government orders. But political considerations play an equal role. Foreign aid officials tend to resist attempts to end a bilateral aid relationship, not only because of possible harm to bilateral relations but also because of prestige, and in view of the fact that the modest Dutch influence in international affairs is considered to be served by having such relationships. Involvement in local development efforts gives 'a place at the [negotiating] table', as a minister of development co-operation put it.

Table 8.1 Dutch ODA: breakdown by number of countries and multilateral channels (percentage)

	1970–71	*1980–81*	*1988–89*	*1993*
bilateral ODA, 10 main recipient countries	58.1	43.8	34.5	20.7
bilateral ODA, all other recipients	16.6	32.6	38.0	49.1
multilateral channels	25.3	23.6	27.5	30.2
Total (US$ million) at current prices	209	1,631	2,288	2,526

Source: calculations based on Lensink 1995.

Over time, ODA expenditure as a share of the overall development budget is increasingly being crowded out by pressures from related budget items, a process known as 'pollution' of the development budget. According to calculations by Lensink (1995: 304), the non-ODA part of the development budget increased from 10 per cent in 1975 to 15 per cent in 1985 and 23 per cent in 1993. In recent years, such pressures have been mounting, due in large part to expenditures on refugees applying for asylum in the Netherlands (45 per cent of non-ODA development spending in 1993). It should be noted that this decline in 'pure ODA' (in relative terms) is perfectly in line with the policy of decompartmentalisation pursued since the early 1990s, and to some extent explained by the new orientation in Dutch foreign aid policy in the 1990s. From the viewpoint of policy integration, a target just for traditional development aid makes less sense, although it has not formally been abolished. A new, and presumably higher, target for integrated international co-operation efforts is yet to be set, however. Nor should the integration of development co-operation into other fields of international activity be viewed as just a move towards putting pragmatic self-interest at centre stage – although this is undeniably a factor. Indeed, integration is also reported to lead to institutional 'encroachment' by the development co-operation institutions upon regular 'foreign affairs' issues (Bik and Nieuwenhuis 1997). Within the ministry of foreign affairs, a department shared by the minister of foreign affairs and his/her counterpart of development co-operation, the latter has much more leverage in budgetary terms. What foreign aid may have lost in developmental purity, it seems to have gained in institutional power within The Hague machinery, especially with a seasoned political heavyweight like Pronk at the helm. This has resulted in a *de facto* division of labour in which the minister responsible for development co-operation is actually a 'Minister of Foreign Affairs for the South'. Thus, representatives from the less advanced developing countries (other than the newly industrialised countries, NICs) visiting the Netherlands are routinely referred to the offices of the minister of development co-operation, not the foreign affairs minister. For a visiting head of state from countries such as Mali or Bolivia, a stop at the Royal Palace will be thrown in for good measure, but the minister of foreign affairs – burdened with serious business – remains a remote figure. This is not always appreciated by leaders thus stereotyped as aid recipients first and foremost.

Since its early days, Dutch foreign aid has essentially followed the course set out by major donors and multilateral organisations, this notwithstanding the pioneer status that Dutch development policy makers have sometimes claimed. In particular, World Bank ideas have been very influential (Kruijt and Koonings 1988). This does not deny the various national accents of the Dutch aid programme, determined by domestic political preoccupations and conditions. Although national political and economic interests affect the allocation as well as the institutional arrangements of Dutch aid, the country has, on the whole, lived up to its reputation as a relatively benign donor, as befits a small nation inclined to moralising, little burdened with the strategic interests of major powers.

Mali, a poor and aid-dependent country

Economic characteristics

Mali, like the other Sahelian countries, is among the poorest countries in the world, with a per capita GNP of $270 in 1993 (World Bank 1996) and with one of the lowest rankings in the UNDP human development index: 171 out of 174 (UNDP 1996). Following the severe drought in the early 1970s, the Sahelian countries received important amounts of development assistance. Since the disaster, aid has no longer come solely from the former colonial power, France. Various donors appeared on the Malian aid scene, including Germany, the USA, Canada, the European Development Fund (EDF), the World Bank and UNDP. The aid industry has emerged with its internal rivalries, while Mali tries to play one donor off against the other.

In 1985, after a second devastating drought, development assistance reached a peak of fl.1,262 million, amounting to 30 per cent of GNP. Subsequently, the total amount dropped (see Table 8.2). In recent years, the economy of Mali has not been threatened by major droughts, but economic growth has been very slow or nearly absent. From 1965 to 1980, the annual growth rate averaged 2.1 per cent, while growth was negative between 1980 and 1993. With a population growth of 2.7–3.5 per cent a year, per capita income has decreased steadily. From 1993 to 1996, annual growth percentages of 2.4 per cent were realised (*Plan Annuel SNV 1997* 1996). Most of the countries of the franc bloc, however, showed sudden growth in this period, following years of stagnation. The 50 per cent devaluation the CFA franc (currency unit of the Financial Community of Francophone African Countries) in 1994 undoubtedly explains a large part of it. Thanks to the SAP, support by the IMF and the World Bank (1996–1998: 45,000 million CFA francs or $86 million) macro-economic conditions appear somewhat better than in the past. The data furnished by the World Bank itself, however, are far from bright, indicating a decrease of GNP and GNP per capita between 1990 and 1995 (including the post-devaluation effect)! Aid dependency does not lessen: 83 per cent of public sector investment in the various economic sectors relies on foreign aid (*ibid.*).

Table 8.2 Development assistance (ODA) to Mali, 1975-1995

	Total ODA in US$ (in millions)	Total ODA in fl. (in millions)	Dutch ODA in US$ (in millions)	Dutch ODA in fl. (in millions)	GNP/ head in US$	ODA as % of GNP
1975	119	301	0.2	0.5	90	22
1985	380	1262	15	51	150	32
1990	506	921	36	66	270	21
1995	591*	988**	33**	56**	50*	22*

Sources: Ministry of Foreign Affairs; IOV 1994; World Bank, *World Development Reports* 1976–1997; *World Development Report 1998* (data for 1994); **Ministry of Foreign Affairs, personal communication (data for 1995).

Typically, agriculture is the most important sector, contributing 40–50 per cent to GNP, while industry is the least important at just 9–11 per cent. Services account for the remaining 40–50 per cent. With the exception of irrigated rice production, agriculture is rain-fed and as a result, depends on weather conditions. This is an important fact for the issue of food security, but it also affects export earnings from cotton, the most important export commodity, and cattle. Mining makes an increasing contribution to exports, but the exploitation of gold and other mineral reserves puts large demands on infrastructure investment, one reason why foreign corporations have been invited to take a lead in this industry.

International development aid to Mali: from food aid to structural assistance

1975–1985

Famine relief or food aid to Mali made a cautious start following the droughts of the early 1970s but received an enormous boost during the second drought period in the early 1980s. In contrast to the previous decade, many aid organisations (foreign and national) were present locally, channels of import and distribution had been established, and local support businesses were available. Thanks to this improved organisational capacity, the second drought did not cause as many fatalities, but Malian society at large was arguably more affected. Food aid did fade away some years after the disaster in the 1970s, but in the 1980s it was continued and transformed into 'structural aid'. Food aid was used to provide the Malian government with financial means to restructure the grain market: this entailed the dismantling of the Office de Produits Agricoles du Mali (OPAM), the semi-public grain marketing monopoly, and liberalisation of the grain markets. Around this restructuring programme (PRMC, Programme de Restructuration du Marché Céréalier), donor consultation and co-ordination took shape. Although such co-ordination would appear a sensible way of preventing duplication of efforts and contradictory interventions, it also had the effect of weakening the Malian government's grip on aid expenditure.

Besides food aid, assistance characteristically aimed at large infrastructural investments, with a special focus on rural areas (production-oriented as well as environmental projects), and a sectoral approach (IOV 1995). France, the former colonial power, continued to play a special role in aid to Mali, not only as the largest donor, but also through its direct interventions in support of the CFA franc, as well as in funding government operational expenditure.

1985–1995

In line with the earlier support for the PRMC, assistance for macroeconomic restructuring prevailed throughout these years. In addition to support for structural adjustment measures, funding was made available for a range of 'good governance' reforms, including the decentralisation of Mali's political structure and its democratisation drive, together with the liberalisation programme, which aims at reducing the formerly dominant role of the state in all domains. The IMF, the World Bank and the Mission de la Coopération Française are important actors and policy makers in this respect. This was made clear by the 50 per cent devaluation of the CFA franc imposed in 1994, a measure with far-reaching consequences. In the 1990s, the donors also claim to be giving more attention to the development of the private sector, in view of industrialisation and in urban development (*ibid.*).

Dutch aid to Mali

Relations between the Netherlands and Mali consist predominantly of aid relations. Commercial relations – trade or investment – are very limited. Commercial exports from Mali to the Netherlands were valued at fl.1.5 million and the commerical imports to Mali from the Netherlands fl.20 million in 1991 (*ibid.*). Official Dutch development aid to Mali is predominantly of a bilateral nature. A part of this aid flow is spent on the CILSS and Club du Sahel programmes, of which Mali receives a share. In the period 1975–1992, two-thirds of the bilateral aid was spent on projects, predominantly in rural areas (IOV 1994). Multilateral expenditure in Mali (by international organisations, e.g. the EDF, the Food and Agriculture Organisation, the World Health Organisation, the World Bank) are partly financed by the Netherlands by way of its contribution to these organisations, but also by allocating funds to special programmes that they operate. An example is a contribution of US$11.4 million in 1996 to the WB Mali programme.

Another form of aid to Mali emanates from many dispersed private initiatives, personal contacts, city twinning programmes, and the like. The Dutch semi-public SNV development organisation stimulates and channels part of these initiatives in addition to the implementation of its own projects. At the moment, for instance, three towns in the Netherlands have aid and exchange relationships with towns in Mali, encompassing support for local initiatives. Another example in this vein is the co-operative efforts of Nuon, a Dutch utility firm, with Electricité de France and Energie du Mali to supply twenty villages around Koutiala with electricity.

Changing budgets

In 1996, the Netherlands (with 6 per cent) ranked as Mali's fourth largest bilateral donor country, after the former colonial power, France (26 per cent), the United States (12 per cent) and Germany (7 per cent), and before Canada (5 per cent). Other important donors (for gifts) were the EDF, and (for loans) the World Bank, the African Development Fund (10 per cent), the African Development Bank, and the Caisse Française de Développement (*Plan Annuel SNV 1997* 1996).

Dutch aid to Mali increased from fl.500,000 in 1975 (US$0.2 million) to fl.66 million in 1990 (US$36 million), after which it continued to fluctuate around the fl.60 million mark (1995: fl.56 million; 1996: fl.62million). Although Dutch assistance to Eastern European countries was stepped up after 1992, this appears not to have occurred at the expense of African recipients and does not explain the relative stagnation in the levels of assistance to Mali.

Changing objectives

As mentioned, in the 1970s the focus in development thinking shifted from the technical aspects of economic growth to problems of distribution and poverty. This was reflected in the reallocation of Dutch foreign aid towards poverty eradication programmes and a geographical shift in favour of Africa, until then not a major recipient. Drought-related famine mobilised considerable assistance for the Sahel region, on food aid, medical supplies and transportation. The first African destinations were Upper Volta (now Burkina Faso) and Sudan, followed by Mali. The priorities for the Dutch Sahel programme during the late 1970s and 1980s can be traced back to the ideas of CILSS and the Club du Sahel:

- Increasing food production and food security.
- Struggle against environmental degradation.
- Water supply for people, cattle and agriculture.

It is understandable that the Dutch adopted this existing programme, considering the lack of experience and local knowledge with respect to the Sahelian context, its problems and potential. Two major research projects, financed from the development co-operation budget, contributed to academic and practical knowledge among Dutch as well as Malian researchers, i.e. on the carrying capacity of pastures in the Sahel and on farming systems in southern Mali.

Besides poverty eradication, a second general focus of Dutch foreign aid in the 1980s aimed at economic self-reliance. Probably due to the incongruence with Malian reality, this focus was not named as a formal objective of Dutch aid to Mali. What can be traced behind this objective, however, is perhaps the concept of 'aid ownership': i.e., some measure of control by the Malians over the formulation and implementation of projects and programmes.

In the 1990s, flexibility in Dutch aid allocations was emphasised, and 'spearhead' programmes were introduced in the fields of ecological sustainability, women's rights, development research and urban poverty. The first two subject areas were already part of Dutch aid policy with respect to Mali (although the term was 'environment' instead of 'sustainability'). They were combined with themes such as integrated rural development, macro-economic support, donor co-ordination, democratisation, human rights and – with a low profile – population policy. In line with the decompartmentalisation strategy calling for new areas of concern such as the internal conflicts in developing countries (Lensink 1995), assistance was further extended to supporting the peace process in the north of Mali in 1996. In order to reconcile warring Touareg and sedentary groups, an integrated regional development plan was prepared to strengthen the war-ravaged economy of the Ménaka area in a joint effort by Dutch organisations and other donors (UN, Red Cross, German Co-operation). Thus, the decompartmentalisation policy of blending development aid with other foreign affairs objectives caused a sudden departure from the long-held policy of concentrating aid geographically in a few key regions in the south of Mali (*viz*. Sikasso and Ségou).

The objectives of Dutch foreign aid may have changed, but these modifications should not be exaggerated. Zoomers (1997) points out that changes are often confined to terminology rather than practice. Dutch objectives, moreover, were influenced by the priorities of the CILSS and the Club du Sahel in the first decades. In the last decade, ideas originating from the World Bank and the IMF are recognisable in the general objectives of Dutch aid to Mali.

Changing geographical focus

From 1975 on, Dutch aid has been flowing to Mali; after 1984, this involved larger amounts. Together with the other CILSS-countries of Burkina Faso, Guinea-Bissau and the Cape Verde Islands, Mali is one of the concentration countries of Dutch aid, a concept introduced in 1976 (IOV 1994: 1). Conditions for this kind of status are observing human rights, continuity in the relationship, and the existence of historical relations. On this basis, it is difficult to grasp why Mali achieved such a position.

Within Mali, the geographical distribution of aid changed over time. After the initial period of famine relief, aid disbursements were gradually concentrated in the southern part of Mali. From 1982 on, two large projects received the lion's share: ARPON, focused on the rehabilitation of the Office du Niger (a major rice irrigation scheme); and DRSPR (Direction de Recherche sur les Systèmes de Production), focused on the development of more productive farming systems in the cotton-growing zone of Mali (*ibid*.). Both involve areas of proven agricultural potential that set them apart as core economic regions within the country. The northern and western regions, with scant economic opportunities, were practically abandoned by Dutch aid efforts, except for small-scale activities of NGOs with an essentially humanitarian orientation. This concentration of efforts on areas of relatively high potential did not easily match the objectives of poverty eradication and redistribution of growth. Management rationalisation was advanced as an argument in favour of spatial concentration (*ibid*.), but economic rationales, such as increasing food production and maximising economic growth, have been mentioned as well (IOV 1995). This suggests that maximising economic impact has been an important consideration in steering Dutch allocations for Mali, even if only partially disclosed. For political reasons in line with the decompartmentalisation of foreign policy, this concentration on southern Mali has been partially abandoned in recent years, as mentioned above.

The large majority of the Malian people (75 per cent) live in the countryside, and in the national budget, the rural economy receives most investment allocations. By far the larger part of development co-operation funds in general, and also those from Dutch sources (90 per cent), are directed to the rural areas. This matches the specifically rural objectives already mentioned, such as increasing food production and food security, the struggle against environmental degradation, and integrated rural development. Yet objectives that are not intrinsically rural in nature also tend to be pursued in a rural context, such as, for instance, a 'women in development' project in rural south Mali.

It is only recently that the urban economy and living conditions have been gaining more attention within the Dutch aid programme for Mali. This is a somewhat belated recognition of the fact that, first, the urban population has been growing rapidly at 6.5 per cent per year in the 1990s (Mission Française de Coopération et d'Action Culturelle au Mali a.o. 1996), with a considerable concentration in the capital city, Bamako, at over a million inhabitants. Second, that the structural adjustment measures have *particularly* affected the urban poor, who tended to rely on public sector spending more than their rural counterparts. Thus, urban poverty alleviation has been added to the Dutch development objectives in recent years, with a key role assigned to NGOs. The semi-public SNV Development Organisation has identified the urban poor, especially women and youth, as a target group (*Plan Annuel SNV 1997* 1996).

Changing channels

A range of channels and aid modalities have been used as a conduit for Dutch aid flows to Mali, and these have all changed over time. Three important modifications have taken place over the last decade that deserve to be mentioned here. They concern (1) the management and administration of assistance, (2) the organisations through which the aid flow is handled, and (3) the mix of aid modalities used.

Until 1988, co-operation with Mali was administered through the Dutch embassy in Dakar in neighbouring Senegal. The growing importance of the Mali aid programme then necessitated the presence of an embassy in Bamako, which was opened in 1990. Following the decentralisation of Dutch development aid in the 1990s, much of the responsibility for programme and project formulation, allocation of the country budget, recruitment of staff and management and control of the country programme has devolved from The Hague to the embassies. The Bamako embassy is a real 'aid embassy', much more actively involved in programme management than could be done at a distance. To perform this expanded role, sector specialists in subject areas such as rural development, gender issues and urban poverty alleviation have been assigned to the Bamako embassy. In addition, Malian nationals are now recruited into the embassy service.

Even before decentralisation, a deliberate policy change in the 1980s sought to rely less on government-to-government co-operation by channeling more aid money through NGOs and MFOs (mede financierings organisaties: co-financing organisations), often based on religious or political identities. Their objectives, policies, target groups and geographical intervention zones are not directly determined by the state, although the ministry retains an important influence through its budget allocation decisions. This is especially true for SNV, a development organisation controlled by the minister for development co-operation with respect to budget and staff, which originated as a volunteer service and is now a professionalised organisation in the field of social development. The share of Dutch aid to Mali spent through such organisations increased from almost 9 per cent in 1986–1990 to 13 per cent of the total in 1991–1995. With their relatively small budgets, they often support small-scale projects. These projects are not restricted to south Mali or the Office du Niger area, where the main bilateral efforts are concentrated.

In terms of aid modalities, project aid and macro-economic assistance have both always been important in the Netherlands–Mali aid relationship. Project aid is restricted in time and space and serves a specific objective. Just as Dutch aid in general is spread over a considerable number of destinations and uses – in spite of repeated calls for a more rational concentration of efforts – so Dutch aid to Mali is scattered over a wide range of activities of varying scope and scale: in 1992, there were thirty-three different projects and programmes. Macro-economic aid, in contrast, takes the form of longer-term intervention in support of a broader goal, and may include food aid, balance of payments support, support for grain market liberalisation, and debt relief. Macro-economic aid was always important in the Netherlands–Mali aid relationship, accounting for 40 per cent of total aid disbursements until 1993, and 56 per cent in 1996. In recent years, there has been a growing emphasis on macro-economic aid and what is called programme aid. The preference for macro-economic aid is somewhat puzzling. It is substantiated by the argument that the results of project aid are often disappointing. However, official evaluations from the ministry's Operations Review Unit are rather ambiguous about the impact of macro-economic aid to Mali. Control of the counter value funds created was almost absent, it was argued (IOV 1995), while the major long-term projects such as ARPON and DRSPR, which took the lion's share of Dutch aid commitments, were in general evaluated favourably (*ibid.*) It is true that evaluations of other, smaller projects ranged from positive to highly negative, but it is unclear how these evaluations would induce a shift in aid modality in favour of macro-economic aid, the impact of which is largely unknown. The expectation that macro-economic aid will yield an improved result seems premature, to say the least (Zoomers 1997). One wonders whether other motives, possibly ease and cost of aid management, have affected this change of preference.

Another new favourite, as mentioned, is the modality of programme aid. Use of the term programme aid is confusing, since in the past, macro-economic aid was also sometimes called programme aid. More recently, it has referred to sizeable clusters of projects for which a lump sum is made available to a Malian recipient organisation, which bears full financial and management responsibility, including the recruitment of technical assistance. In this way, the objective of improving local aid ownership is to be realised. Performance indicators to measure progress are determined on a yearly basis between the Dutch embassy and the Malian implementing organisation. Programme aid enables the Mali government and local organisations to assume more control over development efforts, and facilitates the deployment of national and other (but not necessarily Dutch) specialists.

Evaluation of Dutch aid to Mali

This section will deal with the question of how Dutch aid efforts relate to Mali's 'development'. The first two subsections will discuss some of the achievements and weaknesses of Dutch aid in Mali. Then, some thoughts will be devoted to the influence of the Netherlands on Mali through co-operation.

Achievements

It is impossible to measure the total impact of the Netherlands' aid to Mali. The questions 'did Dutch aid contribute to the development of Mali, and to what extent?' cannot be answered with precision. There are several reasons for this. In the first place, there is no consensus as to the meaning of the concept of 'development' (see Simon, Chapter 2, this volume). Hence, the outcome of any evaluation of development efforts depends on the criteria used to measure this 'development'. Interestingly, major Dutch policy papers such as *A World of Difference* (Ministerie van Buitenlandse Zaken 1990) and *A World in Dispute* (Ministerie van Buitenlandse Zaken 1993) do not contain a clear definition of the concept of development. Moreover, development co-operation serves more aims than 'development' alone. Aid is also given for political, strategic and humanitarian reasons, resulting in aid flows with limited if any developmental impact. The question is further complicated by the fact that the Netherlands is often not the only donor in a programme or project in Mali. In such cases, its contribution cannot easily be separated from the context. Finally, aid has an impact at various levels of society, and may still affect the course of things long after the intervention period. Therefore, we agree with Cassen's observation in his book *Does Aid Work?* (1994) that in the end the precise degree of effectiveness of aid cannot be known. This conclusion also applies to the contribution of aid to 'development'.

Following Otto (1997), we will consider development as an umbrella concept covering a series of interrelated processes, including economic growth, security, equitable distribution of access to resources, conservation of a healthy environment, dissemination of knowledge and power, nation building and good governance, increasing basic health care and cultural self-reliance, all geared to increasing the well-being of all people in society. Based on this concept of development, Dutch aid may be expected to have contributed to particular aspects of Mali's development process. Of course, not all projects and activities turned out equally successfully. The history of Dutch development co-operation with Mali has had failures, such as the construction of a potato warehouse in Kayes that has never been used since completion. Other initiatives, in particular those reflecting the development priorities and approach of Mali itself, proved successful and sustainable (IOV 1995). Overall, the Netherlands, an important donor for Mali, contributed 7 per cent of all aid receipts over the period 1976–1992. Since aid accounted for at least 20 per cent of GNP every year, the conclusion is justified that Dutch aid helped to finance a tiny part of Mali's public expenditure over a sustained period, and assisted the economy through a difficult period of recurring droughts and economic deterioration.

Among the successes in which Dutch aid has had a part have been the famine relief efforts. Beyond mere food aid, in some regions (the diocese of San and district of Koulikoro) small village-level warehouses for cereal stocks have been built (IOV 1994). The amelioration of infrastructure and input supplies to increase food production, fields in which Dutch aid was focused, have also contributed to improved food security. The already mentioned ARPON project contributed to a tripling of rice production per hectare in the period 1981–1991 in the Office du Niger region (IOV 1994). The DRSPR project in Mali-Sud contributed to a better understanding of the existing production systems and, together with the cotton company, developed new methods of sustainable land use (IOV 1994). Mali has also made modest progress in human resource development and institutional capacity building, supported by the Netherlands – among others – through an extensive programme of basic health-care infrastructure in the Ségou region, and through assistance to the Co-ordination Committee for NGO activities (CCA/ONG).

The effectiveness of Dutch aid to Mali improved over time, with the institutionalisation of appraisal and implementation procedures linked to the increasing professionalism of DGIS (General Directorate of International Co-operation). Similarly, improved co-ordination between donor and recipient government and among donors materialised. The Netherlands and Mali conduct policy consultations twice yearly, a high frequency compared with other donors and their suitable channel for The Hague to appreciate the opinions and priorities of the Mali government. It is felt that co-operation between the Dutch and their Malian counterparts has gradually improved, while Dutch co-operation staff, with increasing experience, learned to take Malian attitudes and preferences better into account. Over the years, contacts between Mali and the Netherlands have multiplied. No longer are they confined to exchanges between Malian civil servants and scientists on the one hand and members of the Dutch aid community on the other; nowadays, many other sections of both societies are involved in exchanges that contribute to mutual comprehension. Dutch society at large has gained more regular exposure to Malian art and culture, including aspects of daily life in Mali, through exhibitions in museums and cultural events. In these respects, the long duration of the aid relationship between the Netherlands and Mali has contributed to more knowledge and a better understanding of Mali and its people in the Netherlands.

Progress was made with regard to the objective of concentrating the aid activities in certain regions and sectors. The concentration on the southern Mali cotton area and the Office du Niger irrigation scheme fitted Mali's development plans well, as actually most Dutch aid projects did. From the start, most Dutch assistance was aimed at increasing agricultural production and improving rural production conditions, with a view to attaining food security and improving productivity, especially with respect to cotton, Mali's most important export crop. Within the same context, initiatives to stop soil degradation and to extend rural infrastructure were in line with the development priorities of the Mali government as well as most of the ODA from other donors. In Mali, for obvious reasons, aid emphasised rural development during the first decade, and there was general agreement between Mali's five-year plans, state development investments and external aid. From the mid-1970s on, all major efforts were directed at the overriding objective of securing food self-sufficiency within a stable ecosystem (Cassen and Associates 1994). During the most recent decade, all actors to some degree shifted to activities related to the economic restructuring process, which once again provided a general focus.

Shortcomings

Conversely, weaknesses of the Netherlands' aid effort have to be mentioned as well. Some of these concern the procedural side of aid. In an evaluation, DGIS's Operations Review Unit concluded that the ambitions and requirements of Dutch efforts in Mali increased considerably over time. Along with rising ambitions, procedures became more elaborate. On the one hand, this contributed to increased effectiveness in project design and management, as mentioned above. But on the other hand, increasingly bureaucratic procedures caused long delays between project formulation and implementation. Added to this, projects were (partly) paid from multiple funds, some of which were earmarked for Mali, while others were made available on an *ad hoc* basis. Disbursements often lagged behind commitments, especially when institutional responsibilities between different sections of DGIS were not clear. The fund allocation criteria were often unclear for the recipient, as were the criteria for continuation or termination of projects. An example to this effect is the Dutch assistance for small irrigation schemes in Mali's Mopti region. The Netherlands undertook in principle to finance four small schemes but withdrew its support halfway through. While in the eyes of the Malians the results of the small irrigation schemes were satisfactory, the project evaluation pointed out that the irrigation works would be economically unsustainable as a result of under-utilisation. It turned out that the rice farmers used the schemes only in the dry season, allocating their labour and capital during the rainy season to fields that under 'normal' circumstances would be watered by the Niger River (Harts-Broekhuis and de Jong 1993).

More generally, it is difficult for most Malian counterparts to grasp the complicated financial procedures of the donor. In the early 1990s, for instance, Dutch bilateral aid to Mali was financed through sixteen different funds. These complicated financial arrangements affected the continuity of aid efforts negatively.

Apart from procedural shortcomings, a major problem in respect of aid in Mali is the country's aid dependency, expressed by the enormous share of aid in the overall economy. Aid has in fact become institutionalised as a key sector of the national economy. Aid dependency prevails at two levels: that of the state and national economy, and that of the institutions involved. Without the contributions from its many donors, Mali's government could not function, or at least would have extremely limited scope to undertake development initiatives. Besides, government officials have to invest much effort and time in negotiations with donors in order to defend Mali's interests and to secure the constant flow of aid funds to sustain their activities. Particularly in the case of increased donor co-ordination, this state of affairs could seriously weaken the position of the recipient government. The Netherlands has made an important point of enhancing donor co-ordination.

Aid dependency is also a threat to the institutions that depend on foreign donor funds. From local NGOs to key public organisations such as the IER (Institute of Rural Economics), an important counterpart in many interventions financed by the Netherlands), many institutions owe their very existence to aid and have to extend part of their efforts to sustaining the aid flow. Aid dependency thus incites something reminiscent of what Anne Krueger (1974) called 'rent-seeking behaviour', in the sense that much effort and initiative is spent, not on creating local solutions, but on soliciting external means for solving local problems.

Seen from a development perspective, a major objection against the prominence of donor funding is the lack of coherence between aid efforts. Despite a certain common ground, even the aid activities in Mali of a single donor such as the Netherlands does not constitute an integrated approach to the country's problems. Most development efforts could be characterised as standing in isolation from each other – in terms of location, institutional context and stated objectives. This is a result of the way in which the aid industry works. Individual institutions, consisting of a Malian and a Dutch counterpart (sometimes DGIS itself), formulate project proposals that the Malian and Dutch governments subsequently appraise. The appraisal procedures are based on several objectives laid down in national policy documents, but never derived from an integrated (regional) development plan (see also OECD–UNDP 1998).

Dutch influence on aid to Mali

Even with the best intentions to respect the recipient's sovereignty, aid provides donors with often considerable influence on the country involved. This is partly implicit in aid procedures (see also Hanlon 1991). For instance, until the early 1990s, the core of Dutch aid to Mali was formed by an annual allocation under the 'country programme'. In addition, Mali received considerable aid under various 'special programmes' not tied to a particular recipient. These special programmes allowed the Dutch government to foster themes and sectors to which Dutch policy assigned a high priority. Special programmes were also considered important to raise domestic support for development co-operation (IOV 1995). However, the emphasis on special programmes has varied enormously. This lack of continuity detracted from the sustainability of special programmes assistance. It also caused development efforts in a recipient country such as Mali to be heavily influenced by policy preferences and fashions in donor societies such as the Netherlands.

The donor country also requires control over the use of its aid allocations, partly to ensure adequate alignment with its own policy priorities, and partly in view of financial accountability. Project aid is a much more suitable aid modality for meeting these requirements, irrespective of the question of whether it is the best way of realising a development objective (IOV 1995). Depending on the policy priorities at the time of the budget allocations, the relative amounts of programme and project aid granted varied annually. These fluctuations had little to do with the needs and priorities of the recipient country, which is nevertheless much affected by the outcome. Aid procedures are not only difficult to understand for recipients but also almost completely controlled by the donor.

Until recently, the Netherlands kept substantial control over all stages of the aid project cycle. Already in the preparatory stage, projects tend to be formulated by Dutch professionals. In principle, the project should have emerged from the Malian community, but in practice Dutch experts were much better positioned to tackle the complex criteria and procedures. Subsequently, appraisal of proposals was naturally a Dutch affair as an element of the budgeting process. During the implementation stage, technical assistance tended largely to be retained in the hands of the Dutch. This produced the remarkable result that a larger share of disbursements was retained in the Netherlands in the case of project aid, even though the funds were completely untied (i.e. no conditions were attached on where expenditure could be effected), than was recorded for expenditure under the financial aid procedure, which did carry such restrictions (IOV 1995). Over the last few years, some progress has been made in reducing the control by Dutch nationals. An agreement on programme aid was signed with the cotton development organisation CMDT (Compagnie Malienne de Développement de Textile) in southern Mali, which makes the Malians responsible for all decisions, while the Netherlands will limit its role to supplying funds and after a few years, evaluating whether the objectives, formulated by the Malians, have been met. If successful, this approach could reduce the problem of unwarranted donor influence on recipient countries.

Mali is one of the world's poorest countries. This fact should facilitate meeting one of the priorities of Dutch aid policy, namely that aid be spent to alleviate poverty. However, the concentration of Dutch aid to the Office du Niger irrigation scheme and the cotton-growing area of southern Mali – both areas of high agricultural potential in comparison with other parts of Mali – indicates that Mali's poorest regions and population groups are not the focus of Netherlands–Mali development efforts. One can argue that poverty alleviation does not necessarily mean targeting the poorest among the poor, and that a better result is achieved in terms of economic growth by focusing on regions with relatively good development potential. In the end, this may contribute more to sustainable development and well-being than channelling investment into projects and areas with poor prospects.

The concentration of development investment in *le Mali utile* (useful Mali) has another side. It contributes to geographical polarisation in economic terms and hence to regional inequality within Mali. Thus, aid is a factor in the increased migration flows from poor, remote northern regions of the Malian Sahel to the area of the Office du Niger and the southern cotton belt. This influx of migrants raises population pressures in the south, and hence the risk of ecological degradation. It is difficult to resist a migrant's wish to participate in southern development projects. In the Office du Niger irrigation scheme, it is proposed to reduce the size of family land holdings with a view to increasing the number of beneficiaries. Implementation of this proposal will certainly lower the viability of the farms in the Office, a key supplier of food (rice) for Mali. Development policy has to make a precarious trade-off between humanitarian concerns – enabling people to survive at subsistence level – on the one hand and the creation of a viable economy on the other. Mali needs both, and cannot do without aid in support of the one or the other.

Conclusions

Development thinking certainly influences the practice of development policy and aid, but not in a simple, direct way. For one thing, development is practice often lags behind development thinking. Also, although general trends can be detected, thinking on development is diverse, and varies with interest, ideological stance and so forth. Ideas change in response to changing circumstances in the Third World and as a consequence of feedback processes. Systematic feedback, however, appears not to be a strong element of development aid.

Development thinking is intrinsically idealistic. The objective of development is improving conditions, whether this is cast in terms of economic growth, modernisation, human development or something else. As a natural corollary, development aid is seen as 'doing good', and this is the basis of public support for aid efforts. The point of departure in development aid is often that the direction development should take is known. In fact, the knowledge on how to improve things is limited and is often based on experiences elsewhere and in the past. The dependency and imperialism paradigm attempted to show that developing countries have a different position in the world economy, and that internal economic processes are shaped by this position, challenging the models from the core economies. The view that the development paths of individual developing countries will differ from the those pursued by the 'developed' countries is now fairly widespread. The rather blueprint-like approach favoured by the World Bank and others, stressing similar structural adjustment processes in widely different countries, seems to contradict this, and many of the critiques of adjustment policies indeed stem from doubts about a single ideal development model (Simon 1995).

Although development aid moved away from handing over capital and goods to a transfer of technology and knowledge, this transfer from donor to recipient countries at heart still reflects the view of the transferability of development experience following an implicit unilinear model. In fact, the 'democratisation' of aid, through which sections of society with limited theoretical knowledge of development are involved in development co-operation (e.g. NGOs, town twinning, professional and trade organisations) may strengthen the unilinear conception of development. Postmodernism, which implies acceptance of differences in conditions and orientations, as yet has little influence on the practice of official development co-operation (see Chapter 2).

These implicit ideas on the transferability of development experiences can also be retraced to the institutional sphere. Bilateral co-operation implicitly assigns a certain role in development to the nation-state. In many Third World countries, the state is often a relatively recent phenomenon, inherited from colonial times. Governments have not often been democratically elected and, in any case, represent the interests of certain groups at the expense of others. Official development aid contributes to maintaining this situation. Democratisation and the pursuit of 'good governance', pet subjects in recent development co-operation, may be desirable but at the same time can be as strange to local experiences as the concept of the nation-state. The Dutch preference for promoting democratisation processes in Mali similarly does not address the question of whether a Western-style democratic constitution is the right answer for the country in question. The development experiences of Southeast Asia (Singapore, Malaysia) teaches us that other political arrangements are also possible and may actually have advantages in pursuing (economic) development.

Aid flows may become very important for the survival of the recipient institutions. In order to secure aid flows, the Malian government, as well as NGOs, have felt compelled to adopt 'donor development values' as their own objectives. This may serve the purpose of securing necessary assistance, but it also reinforces the dependency of recipient countries. In view of the different and sometimes unfortunate role of the state in Third World countries, it is astounding that almost all aid has been geared to supporting the public sector, while only a small share has been directed to non-governmental agents. In the 1990s, larger amounts of aid are being channelled to private actors, non-profit organisations and local development agencies. Initiatives deployed by private companies are still rare and viewed with suspicion.

In order to maintain public support, systematic evaluation of development assistance and periodic scrutiny of expenditure by the Dutch parliament is necessary. However, increasing macro-economic assistance and more reliance on local counterparts in project formulation and implementation will hinder both the appraisal procedures and the financial accounting options for the Dutch parliament. It is questionable whether aid allocated on contractual terms, as was done with the cotton development agency, will solve this problem. Decentralisation of the appraisal of aid activities to the embassies is in conflict with the administrative logic of accountability and political responsibility with respect to the Dutch parliament.

Although public support for development aid is based on the ideal of promoting development for the Third World, Dutch aid is now determined more than before by policy considerations other than just 'development'. The decompartmentalisation process implemented in the Dutch foreign ministry underscores this trend. Development co-operation policies serve many hidden agendas, which can often be understood only (long) after the implementation of the policy.

References

Bik, J.M. and **Nieuwenhuis, W.H.** (1997) Wie betaalt, blijft bepalen, *NRC Handelsblad* (29/5/1997: 33).

Cassen, R. and Associates (1994) *Does Aid Work? Report to an Intergovernmental Task Force* (2nd edn). Clarendon Press, Oxford.

Cooper, C. and **van Themaat, T.** (1989) Dutch aid determinant 1974–85: continuity and change. In Stokke, O. (ed.) *Western Middle Powers and Global Poverty. The Determinants of the Aid Policies of Canada, Denmark, the Netherlands, Norway and Sweden,* The Scandinavian Institute for African Studies, Uppsala.

Hanlon, J. (1991) *Mozambique: Who Calls the Shots?* James Currey, London.

Harts-Broekhuis, E.J.A. and **de Jong, A.A.** (1993) *Subsistence and Survival in the Sahel. Responses of Households and Enterprises to Deteriorating Conditions, and Development Policy in the Mopti Region of Mali.* Nederlandse Geografische Studies, 168, Utrecht.

Hoebink, P. (1997) Theorising intervention: development theory and development co-operation in a context of globalisation. In van Naerssen, T., Rutten, M. and Zoomers, A. (eds) *The Diversity of Development,* Van Gorcum, Assen, 362–372.

IOV (1994) *Evaluatie van de Nederlandse hulp aan Mali 1975–1992* Inspectie Ontwikkelingssamenwerking te Velde, Ministerie van Buitenlandse Zaken, Den Haag.

IOV (1995): *Evaluation of Netherlands Aid to India, Mali and Tanzania.* Netherlands Development Co-operation, Ministry of Foreign Affairs, The Hague.

Krueger, A. (1974) The political economy of the rent-seeking society, *American Economic Review* (June).

Kruijt, D. and **Köonings, K.** (eds) (1988) *Ontwikkelingsvraagstukken: theorie, beleid en methoden.* Coutinho, Muidenberg.

Lensink, R. (1995) Decompartmentalisation: the dismantling of Dutch development co-operation? *Tijdschrift voor Economische en Sociale Geografie* 86(2): 303–308.

Lieten, K. and **van der Velden, F.** (1997): Over de grenzen van de hulp. In Lieten, K. and van der Velden, F. (eds) *Grenzen aan de hulp, beleid en effecten van ontwikkelingssamenwerking,* Het Spinhuis, Amsterdam.

Ministerie van Buitenlandse Zaken (1990) *Een Wereld van Verschil, nieuwe kaders voor ontwikkelingssamenwerking in de jaren negentig.* SDU, Den Haag.

Ministerie van Buitenlandse Zaken (1993) *Een Wereld in Geschil. De grenzen van de ontwikkelingssamenwerking verkend.* SDU, Den Haag.

Mission Française de Coopération et d'Action Culturelle au Mali, Club du Sahel and **CEPAG** (1996) *Le Mali dans le XXIe siècle. Actes du séminaire Perpectives à long terme en Afrique de l'ouest et au Mali: conséquences pour la coopération.* Bamako.

OECD–UNDP (1998) *Review of the International Aid System in Mali, Synthesis and Analysis, Provisional Report,* Internet (27/2/1998).

Otto, J.M. (1997) Ontwikkelingssamenwerking en goed bestuur, *Internationale Spectator* **51**(4): 223–229.

Plan Annuel SNV 1997 (1996) SNV, Den Haag.

Preston, P.W. (1996) *Development Theory: An Introduction.* Blackwell, Oxford.

Schrijvers, J. (1997) Ontwikkelingsparadigma's. Een kritisch perspectief. In Lieten, K. and van der Velden, F. (eds) *Grenzen aan de hulp, beleid en effecten van ontwikkelingssamenwerking,* Het Spinhuis, Amsterdam.

Simon, D. (1995) Debt, democracy and development: sub-Saharan Africa in the 1990s. In Simon, D., van Spengen, W., Dixon, C. and Närman, A. (eds) *Structurally Adjusted Africa: Poverty, Debt and Basic Needs,* Pluto, London.

Sterkenberg, J.J. and **van der Wiel, A.** (1995) Structural adjustment, sugar sector development and Netherlands aid to Tanzania. In Simon D., van Spengen, W., Dixon, C. and Närman, A. (eds) *Structurally Adjusted Africa: Poverty, Debt and Basic Needs,* Pluto, London.

UNDP (1996) *Human Development Report 1996.* Oxford University Press, New York.

World Bank (1996) *World Development Report 1996.* Oxford University Press, New York.

Zanen, S. (1996) *Ontwikkeling: idee en handeling.* CNWS, Leiden.

Zoomers, A. (1997) De moeizame weg naar een nieuw ontwikkelingsbeleid, *Internationale Spectator* **51**(4): 218–222.

CHAPTER 9

The Pacific Asian challenge to neoliberalism

Chris Dixon

Introduction

The Pacific Asian economies have come to occupy a critical position in the development debate. Their development since the 1950s has been the subject of a voluminous literature, which has generally divided sharply on ideological grounds. On the one hand, there are those who have depicted the emergence of the four Asian NIEs (the newly industrialising economies of Hong Kong, Taiwan, Singapore and South Korea) as the result of open, free-market development where 'State intervention is largely absent. What the state has provided is a suitable environment for entrepreneurs to perform their functions' (Chen 1979: 84). In contrast, a large number of writers (for example, Amsden 1989; Castells *et al.* 1990; Wade 1990; White and Wade 1988) have concluded that the growth of NIEs was the product of extensive state intervention. Indeed, White and Wade (1988) described the NIEs as 'guided capitalist markets', drawing attention to the similarities with the Asian centrally planned economies. Despite differences in ideology and economic systems between the East Asian countries:

> the role of the state has been prominent in all. ... In each case the developmental impact of the state extended far beyond economic policy to include ideological mobilisation, pervasive political controls and social engineering. Each state has sought to define and implement economic priorities through varying forms of strategic planning.
>
> (*ibid.*: 24)

Despite the weight of material that has emphasised the central role played by the state in Pacific Asian development, this has generally not been accepted by the international agencies. Indeed, they have usually followed the line proposed by Friedman and Friedman in 1980, stressing the contrast between the dynamic market economies in Pacific Asia and the stagnation of those that were centrally planned. This view has become a cornerstone of the current neoliberal development orthodoxy promoted by the international agencies and enshrined in structural adjustment programmes. Indeed, the successful Pacific Asian economies have been presented as living embodiments of neoliberal orthodoxy and examples to be followed by the Third World as a whole. The debate over the development of the Pacific Asian economies has intensified since the mid-1980s with the apparent spreading of miracle growth into Indonesia, Malaysia and Thailand and, during the 1990s, into the PRC (People's Republic of China) and Vietnam. A further dimension to the debate was added during 1997–98 by the rapid emergence of a series of economic crises, which have particularly affected Indonesia, South Korea and Thailand.

It is the purpose of this chapter to review the conflicting interpretations of the development of the Pacific Asian economies in the light of events since the early 1980s and outline the implications for the theory and practice of development.

The Pacific Asian regional pattern of growth

The development of the Pacific Asian economies can be seen in terms of the interaction of particular national, regional and international conditions that have enabled the emergence of four generations of industrial development. These comprise Japan during the 1950s, the NIEs (Hong Kong, Singapore, South Korea and Taiwan) during the 1960s and 1970s, the ASEAN-3 (Association of South East Asian Nations – Indonesia, Malaysia and Thailand) during the 1980s and the ATEs (Asian Transitional Economies), particularly the PRC and Vietnam, during the 1990s. While all these stages were characterised by the rapid development of manufacturing and export earnings, there were profound differences in the nature and historical experience of the states concerned and the prevailing international and regional circumstances. However, there has been a tendency to treat the emergence of the Pacific Asian economies, particularly the second and third generations, as the product of a series of common factors that together provide a model for development elsewhere in the Third World. This view has been reinforced by the manner in which the spreading of rapid export-oriented growth has been accompanied by increasing linkages between countries.

By the mid-1990s, the Pacific Asian region had become remarkably highly integrated, with 55 per cent of trade taking place within the region in 1996. Since the early 1980s, there has been a sharp increase in the proportion of trade and investment that is intra-regional (Dixon and Drakakis-Smith 1995; Dixon 1998b). In response to rising costs, concern over pollution and policies aimed at upgrading production, manufacturing processes have been progressively decanted from Japan and later the NIEs. Large numbers of Japanese and NIE-based companies have gone transnational on a regional basis, relocating activities into the ASEAN-3 and, more recently, the ATEs. This pattern of disaggregating industrial production at the regional level has also been followed by many European and North American companies. The result has been the formation of a nested series of regional divisions of labour, based principally on differences in labour costs (Henderson 1986, 1989; Dixon 1991:

218–222; Dixon and Drakakis-Smith 1995). For some observers (see for example Kwan 1994), the economic dynamism and increasing integration of the Pacific Asian region suggested that it had broken the links with Europe and North America and could generate long-term economic growth, at least in part, independently of global trends. The rapid spreading of the 1997–98 financial crises (see below) confirms the level of integration but brings into question the degree to which Pacific Asia has become independent of global forces. The heavy dependence on regional markets, which has increasingly shielded the Pacific Asian economies from recession and protectionism in Europe and North America, has made it very difficult for countries such as Thailand and South Korea to follow the IMF suggestion that they export themselves out of their crises.

The neoliberal interpretation of the Pacific Asian growth economies

The neoliberal 'counter revolution' in development theory (Toye 1987) emerged from the debt-related crises that affected the majority of Third World economies during the early 1980s. Within Pacific Asia, the limited and generally short-lived impact of the crises on the region's economies, coming on top of their sustained rapid growth and structural change since the 1960s, attracted considerable attention. The success of the export-oriented Pacific Asian NIEs (Hong Kong, Singapore, South Korea, and Taiwan) was seen as standing in marked contrast to the crisis-ridden import-substituting economies. This gave additional weight to the international agencies' disenchantment with import substitution-based development strategies, a view which had been emerging since the early 1970s. The Pacific Asian NIEs were depicted not as exceptions or special cases within the Third World but rather as the product of the adoption of export-oriented strategies, behind which lay 'sensible' internal policies based on 'sound neoclassical principles' (Tsiang and Wu 1985: 329). Out of this, the view began to emerge that the Pacific Asian economies might be presented as a 'model' of development for the Third World economies as a whole. The resultant internal policy-based model had three components: limited government intervention in the economy, low levels of price distortion, and an outward-oriented strategy of export promotion. While the details of policy and, indeed, the exact meaning of the key terms remained far from clearly stated and prone to change over time (see in particular Gore 1996), they became central to the neoliberal prescription for the Third World as a whole and basic to the structural adjustment programmes implemented from the early 1980s onwards.

During the 1980s, the three components of the neoliberal model were collapsed by the agencies into one simple requirement to 'get prices right'. It was asserted that there was a close relationship between countries with low levels of 'price distortion', outward orientation and high levels of economic growth (World Bank 1983). The empirical basis for this view has been heavily criticised in terms of the data, unsatisfactory and inconsistent definitions of 'price distortions' and 'outward orientation', and the direction of the causality – if any (Jenkins 1991; Singer 1988; Gore 1996).

The simple neoliberal interpretation of the Pacific Asian economies that became current during the 1980s has been heavily and extensively criticised (see in particular Amsden 1989; Rodan 1989; White and Wade 1988). Essentially, these writers stressed the extent to which the development of the Pacific Asian NIEs was the product of high levels of state intervention and considerable distortions of prices. Indeed, the success of the Asian NIEs can be attributed to them getting the prices 'wrong' in agency terms (Kiely 1998: 78; Amsden 1989: 139–155). Other criticisms centred on the very special nature of the NIEs.

Hong Kong and Singapore were city states with major *entrepôt* functions and high degrees of nodality within the regional and global economies. South Korea and Taiwan inherited substantial infrastructure and industrial development from the Japanese colonial period, achieved radical social and political change that largely eliminated the power of landed and rural interests, and during the period of the Cold War received large amounts of American aid and technical assistance (see, for example, Cummings 1987). Overall, the rise of the NIEs during the 1970s has to be seen in the context of changing international and regional circumstances and the possibilities offered by the crises of profit in Western manufacturing, the flight of industrial capital, the related internationalisation of production and finance, and the decanting of labour-intensive activities. Under the neoliberal internal policy-centred model, all these factors were treated as 'residuals'.

Since the late 1980s, in response to the volume of criticisms and the increasing prominence of evidence of continuing state intervention in the NIEs, the neoliberal interpretation of their development underwent a measure of change. The revised view is most clearly presented in the World Bank's 1993 study *The East Asian Miracle*. This report examined the growth of what it termed the highly performing Asian economies (HPAEs); these comprise Japan and the four Asian NIEs, together with Indonesia, Malaysia and Thailand. The focus was on internal policy, with little attention devoted to international, historical or other factors. Inevitably, given the extremely varied nature of the HPAEs, the policy prescriptions tended towards the most general common denominators. Not surprisingly, the analysis has again been heavily criticised (see in particular Amsden 1994; Gore 1996; Kiely 1998).

There are, however, some significant differences in the revised neoliberal interpretation of the Pacific Asian experience and the related prescriptions for other Third World countries that have become current during the 1990s. The simple slogan of 'get prices right' has been replaced by a broader-based exhortation to 'get the basics right'. This stresses the need for 'sound macro-economic policy and stability'. Thus the World Bank (1994: 2) has emphasised that 'good macroeconomic policies have paid off in East Asia, and they will pay off in Africa'. However, it is again possible to criticise this focus on broad macro-economic indicators in terms of their selection and the ignoring of long-term imbalances in a number of the Pacific Asian economies that would certainly not conform to IMF or World Bank expectations. Indeed, in this respect, if all the HPAEs had kept to the macro-economic limits now advocated by the international agencies, the World Bank might have had no need to invent the term HPAE.

Unlike the earlier neoliberal interpretation, *The East Asian Miracle* report does allow some role for the regional situation, proximity, culture, history and American aid. However, these are downplayed and dismissed in two pages (out of 389). Indeed, it is suggested that all these individual features can be found in countries that did not become HPAEs. Nothing, however, is said of their combination or interaction with other factors. Effectively, the residual factors are raised only to be dismissed as of negligible importance.

More significant than the broadening of the agencies' key development concept and slogan was a revision of the way in which internal policy and the state were viewed. Here, as Gore (1996: 100–101) has demonstrated, there has been considerable sleight of hand, if not 'double think'. Once it was no longer tenable to maintain the fiction of the non-interventionist Pacific Asian states, it was necessary to subsume the activities of the states into the neoliberal paradigm. This was achieved by describing state intervention in the HPAEs as 'market-friendly'. Under this conceptualisation, the state intervenes reluctantly when there are failures of the market and subjects the intervention 'to the discipline of the international and domestic markets' (World Bank 1991: 5). The intervention in the HPAEs is also seen as replicating the behaviour of the market by the operation of a system of rewards and penalties. Such intervention resulted in a low level of distortion of markets and prices, a marked contrast to the results of state activities elsewhere in the Third World (World Bank 1993: 351). This conclusion appears to be contradicted by data presented in *The East Asian Miracle* report, which suggested that the price distortion in some of the HPAEs was greater than that found in some 'unsuccessful' Latin American states (World Bank 1993: 301). Perhaps more bizarre is the argument that many interventions effectively cancelled one another out. A logic that suggests that the level of state intervention in South Korea and Taiwan was reduced because the protection of the import-substituting sector was balanced by the promotion of exports (World Bank 1993: 305) is difficult to fathom.

The World Bank does, however, find evidence for state interventions that do not conform to the market-friendly image. This is notably the case with respect to heavy industry in South Korea, which is described as costly and inefficient (World Bank 1993: 305). However, as Amsden (1989: 243–318; 1994: 630–631) has demonstrated, South Korean heavy industry was a remarkable success: by the early 1980s, it was highly competitive in international markets and contributed the majority of the country's export earnings.[1] Amsden's (1989) study of South Korea is perhaps the most detailed and persuasive (and widely cited) evidence against the World Bank's position on price distortions.

[1.] Centralised direction of the location of heavy industrial plants was also a major component of South Korea's relatively successful programme of industrial decentralisation (see Auty 1990, 1992).

While the revised neoliberal position recognised the role of the state in the successful Pacific Asian economies, it was still substantially played down. Indeed, it is suggested that there is no evidence that the activities of the state in any way helped development, and perhaps in its absence the economies concerned would have been even more successful (World Bank 1993: 6). However, at no point does the World Bank attempt to investigate the relationship between Pacific Asian economic growth and state intervention; it is merely stated that it is impossible to measure, from which its non-existence is then concluded (*ibid.*: 6). At the same time, the World Bank is apparently quite happy to accept the counter proposition (again with no attempt at testing) that in a large number of cases, state intervention has been inimical to development (*ibid.*: 325). There is here a strong message for other countries that may balk at reforms aimed at establishing market systems and reducing the central role of the state:

> The fact that interventions were an element of some East Asian econo-mies' success does not mean that they should be attempted every-where, nor should they be used as an excuse to resist market-oriented reform.
>
> (Page 1994: 624)

Within the neoliberal framework, there are some who give the role of the state in Pacific Asian development far greater prominence. This is perhaps most clearly stated by Krugan (1994). Here the Pacific Asian growth economies are likened to the USSR during the 1950s and 1960s, rapidly growing on the basis of central direction of the economy and thereby providing a worrying challenge to the Western market economies (*ibid.*: 63). These concerns were to be dispelled by the slowing, stagnating and eventual collapse of the Soviet economy during the 1980s. Thus it is argued that while state intervention and a high degree of centralised control over the economy can mobilise resources and dramatically increase inputs, in the long run high rates of growth cannot be sustained, because only the free market is effective in raising productivity as against merely increasing the level of inputs (*ibid.*). This view does not fit with other evidence of increasing factor productivity within the Pacific Asian economies produced in support of the neoliberal position (see Page and Petri 1993; World Bank 1993). However, it should be noted that there is evidence for a marked slowing in the growth of productivity in a number of Pacific Asian economies during the 1990s (see, for example, *Financial Times*, 29 August 1997, on Malaysia). Overall, the 1997–98 crises have given some credence to the idea that state-led development is not sustain-able in the long run.

The development of the Pacific Asian economies from the 1950s through to the mid-1990s was extremely difficult to fit convincingly into the neoliberal framework (see Wade 1996). However, for the international agencies and the principal Western donors it was thought necessary to maintain the charade in order to present the Pacific Asian economies as the product of free-market, open capitalist regimes, the dynamism of these economies being presented as forming a sharp contrast to the stag-nating Asian 'socialist' bloc. For adherents to this view of the ineffectiveness of central planning, it was reinforced from the early 1980s onwards by the apparent economic stagnation and implementation of 'reforms' in the socialist bloc.

The Asian transitional economies

Since 1979, the governments of Cambodia, Laos, Myanmar (Burma), the PRC and Vietnam have initiated programmes of economic reform; these accelerated from the mid-1980s and, more especially, from the end of the decade. To a degree, the process of reform in these countries is linked, with superficially similar policies having been introduced in response to' broadly comparable problems and international circumstances. Overall, since the late 1980s, policies in all the ATEs have become market-oriented and outward-looking. These policy changes have become associated with marked accelerations in the rate of growth of GDP and changes in the internal and external structures of their economies. In these respects, there are marked similarities with the experience of the earlier genera-tions of Pacific Asian growth economies.

In all the ATEs, the programmes of reform appear to reflect deep-seated (although far from unanimous) changes in economic perspectives on the part of the ruling parties. To a degree, these are reflected in the terms applied to the reforms, namely *doi moi* (new way) in Vietnam and *chin thanakan mai* (new thinking) in Laos. These terms contrast markedly with those of *glasnost* (openness) and *perestroika* (restructuring) popular-ised in Poland and the former Soviet Union respectively, but now so widely, and perhaps inappropriately, applied to transitional economies in general.

The ATEs not only differ markedly from the Eastern European econo-mies but are in themselves by no means a homogeneous group. They differ profoundly in history, culture, society, economy and state form. There were, for example, substantial differences in the degree of centrali-sation and socialisation of the means of production before the implemen-tation of reforms. In these respects, there is probably a continuum that runs down from the PRC through Vietnam, Myanmar (Burma), and Laos to Cambodia. There is here the whole neglected issue of what the transi-tional economies are transitional *from*.

The ATEs are among the world's poorest economies[2] and are in the process of integrating into the world's most dynamic region (the 1997–98 crises – discussed below – notwithstanding). While the ATEs have very low levels of per capita GDP, they have, in some cases, tended to rank rather better in terms of such measures as infant mortality rates, life

[2.] In 1995, the per capita incomes of the ATEs were:

	US$
Cambodia	270
Laos	350
PRC	620
Vietnam	240

Source: World Bank (1997) *World Development Report*. Oxford University Press, New York.

211

expectancy and literacy rates than other low-income countries. This was particularly the case for Vietnam, with its provision of welfare, health and educational facilities and housing. Since 1989, the shortages of funds have reduced the provision, and payments for health and education have been introduced. In combination with the reduction in the level and security of public sector employment, these changes have resulted in marked increases in income disparities and poverty (Dixon and Drakakis-Smith 1997: 29–33). However, the Vietnamese government has stressed that this reduction in welfare provision is temporary. There is no evidence to suggest that either the Vietnamese state or people have come to accept the view that is being promoted in Pacific Asia that welfare is the concern of the individual and the family rather than the state. Indeed, there is considerable evidence to suggest that the Vietnamese people expect the state to take responsibility for a wide series of issues, including the regulation of the actives of foreign firms, wage levels and working conditions (Beresford 1993; Noerlund 1997, 1998). This situation tends to place Vietnam apart from the earlier generations of Pacific Asian growth and at variance with the tenets of neoliberal orthodoxy.[3]

The process of reform in the ATEs differs markedly from the development in Eastern Europe. Most significantly, reforms in the ATEs have taken place on a more gradual and long term basis without major and abrupt political change. Only in Myanmar (Burma) was there the appearance of abrupt change with the change of name from the Socialist Republic of Burma to Myanmar and the repeal in 1989 of the 1965 Law of Establishment of the Socialist Economic System. However, this has been followed by the slowest reform of any of the ATEs (see Myat Thein and Mya Than 1996) and accompanied by political repression. Thus, as Katz (1996: 10) has concluded, the

> Asian nations did not feel the contempt for government and the need to erase overnight decades of state economic intervention which characterised Central and East European reform.

The process of economic reform in the ATEs was not accompanied by the moves towards 'democratisation' that characterised the Eastern European states. Only in Myanmar (Burma) and Cambodia have there been multi-party elections, and in the former they were rendered ineffective by the ruling SLORC (State Law and Order Restoration Council) and in the latter imposed and largely funded from outside, only to be negated by the 1997 *coup d'état*. In Laos, the PRC and Vietnam, centralised one-party rule remains firmly in place, with few signs of any immediate change. Thus in terms of both political and economic reform, the ATEs contrast very markedly with the situation in Eastern Europe.

In the ATEs, reforms have been centrally planned as part of a long-term process within which the state is envisaged as playing a major ongoing role. Again, this contrasts sharply with the Eastern European

---

[3.] In the longer run, concern over welfare might place Vietnam in a disadvantaged position in terms of attracting foreign investment. See, for example, Dragsbaek Schmidt's (1997: 26, 28–30) discussion of the role of 'comparative austerity' in reducing Southeast Asian costs and attracting investment.

'big bang', under which the central controls were removed rapidly, leaving a vacuum that rapidly became occupied by rather truncated versions of the market. As Katz (*ibid.*: 18–19) has noted, while the rapid removal of the state's functions has great ideological appeal, it is contrary to the experience of such countries as the United States, Germany and the United Kingdom, and perhaps even more, such Asian countries as Japan, South Korea and Taiwan. To this has to be added the serious social and economic impact of the lack of adequate regulatory and institutional frameworks in Eastern Europe (*ibid.*: 15–16). Despite this, behind much of the comment on the transitional economies, there appears to be a desire to see the development of some ideal-type market economy that has never existed in the West and even less so in Asia. This represents the extremes of the neoliberal orthodoxy that has prevailed since the early 1980s and that has promoted the reduction of the role of the state as a necessary requirement for economic dynamism.

Despite the experience of the earlier generations of Pacific Asian growth economies, there was an expectation on the part of the international agencies that there would be rapid privatisation of the state sectors in the ATEs.[4] However, unlike Eastern Europe, there has been no wholesale privatisation of SEs (state enterprises) in the ATEs. Only in the limited state sectors of the small, weak Laotian and Cambodian economies have cash-strapped governments undertaken any appreciable privatisation (Livingstone 1997; St John 1997). However, even in these cases, the lack of domestic capital and limited foreign interest has meant that much of the privatisation is highly cosmetic, with the former SEs remaining closely linked to the government and dependent on subsidies and concessions. In Myanmar (Burma), although privatisation has been a policy since 1988, it has had no priority and little has happened in practice. This contrasts with the PRC and Vietnam, where policy has been firmly against privatisation, but the SEs have been subject to reforms that have been aimed at encouraging them to operate increasingly within a market framework. There is no indication that privatisation is on the Beijing agenda, whatever the opinion of the international agencies. Particularly in Vietnam, there has been rapid reform and streamlining of the state sector in conjunction with a commitment that the 'commanding heights' of the economy will remain in state hands (Beresford 1993: 226). In the PRC, the slower pace of SE reform and the continued inefficiency and loss-making capacity of the undertakings continues to be a concern (Sender *et al.* 1997: 70).

In Vietnam, there have been moves since 1979 to increase the autonomy of state enterprises and end the system under which they were allocated inputs and machinery (over the quality of which they had no control) and were required to supply their output to the state trading network. Particularly since 1986, reforms have progressively given the state enterprises responsibility for obtaining inputs and equipment, selling directly to customers, conducting their own market research and invest-

---

4. This has been an ongoing source of friction between, for example, the Vietnamese government and the IMF and World Bank (*Financial Times*, 20 April 1998).

ment. Financial targets are no longer set centrally, and profits and cash holding can be retained rather than deposited with the state bank. The only financial targets set are the contributions to the state budget (Beresford 1993: 226). These developments, together with the ability to acquire funding outside the state sector, have given a high degree of autonomy to the state sector, except for some concerns in the transport and power sectors (*ibid.*: 226). The managers of the enterprises have complete autonomy to employ and sack labour and determine wage levels and conditions, subject to government regulations (*ibid.*: 226). Thus, overall, reforms have been directed at increasing efficiency, profitability and market-oriented behaviour (Grahner and Stark 1996: 1).

While the reform of the Vietnamese public sector has not involved privatisation, it has been accompanied by the closure and merger of concerns, which reduced their number from 12,000 in 1990 to 7,000 by the end of 1993 and nearer to 6,000 by 1997 (St John 1997: 179). There is evidence to suggest that the reforms of the state sector in Vietnam are proving successful. In many cases, enterprises are out-performing the private sector. However, as St John (*ibid.*: 180) has noted, many state enterprises have benefited from the establishment of joint ventures with foreign investors, which have given them access to technology, finance and expertise not readily available to the private sector. In addition to the favourable performance, there is evidence of better working conditions (*Vietnam Labour Watch*, cited in Noerlund 1997: 16). Of course, some enterprises continue to perform poorly (St John 1997: 180) and some managers and officials are unable to adjust to and/or resist the changes (Beresford 1993: 226).

A critical issue in the retention of the state concerns in the ATEs is that they did not come to dominate the economies in the manner associated with Eastern Europe. This is a reflection of the low level of economic development and, particularly in Cambodia, Laos, Myanmar (Burma) and Vietnam, the incomplete socialisation of the means of production. These situations left considerable scope for new private sector activity. In contrast, in Eastern Europe the scale and scope of the SEs meant that their continued existence seriously constrained the development of the private sector. In addition, unlike the ATEs, the major East European SEs were attractive to foreign investors and seen as sources of immediate funds for cash-strapped governments.

Despite the varied situations that have prevailed in the ATEs since the mid-1980s, the progress of reform has been rapid. A critical element in the reform processes has been the rapid opening of the economies to trade and investment and integration into the Pacific Asian regional economy. This has been institutionalised with the extension of ASEAN membership to Vietnam in 1996, Laos and Myanmar (Burma) in 1997 and Cambodia during 1998.

Thus the ATEs are becoming integrated into wider regional economies in which long-term high rates of growth have been closely associated with high levels of state intervention and direct involvement in production. Here there have been none of the hang-ups over such activity on the part of the state that have come to characterise Western views (Dragsbaek Schmidt

1997: 26, 39–40). The situation that is emerging in the ATEs, particularly Vietnam and the PRC, has been labelled 'market socialism', a concept closely linked with White and Wade's (1988) discussion of the emergence of 'socialist-guided markets' with marked similarities to East Asian capitalist-guided markets. These developments, reinforced by increasingly close economic and political linkages with the rest of Pacific Asia, provide a stark contrast to the policy prescriptions and expectations of the international agencies. Thus the development of the ATEs suggests that there are perhaps trajectories for transitional economies and poor Third World countries as a whole that do not conform to the neoliberal framework.

The 1997–98 crises

With the onset of the 1997–98 Pacific Asian crises, the above view of the unsustainability of the form that development had taken was reinforced (for a detailed account of the origins of the crises, see Rosenberger 1997). The IMF and the World Bank have 'discovered' or in some cases 'rediscovered' forms of state intervention and market distortion that, according to such studies as *The East Asian Miracle* report, did not exist. For South Korea, the comments and conditions attached to loans mirror those made during the brief recession and debt crisis of the early 1980s (World Bank 1993: 309; Dixon 1995a: 214).

With the end of the Cold War, there was no longer any need to present such economies as South Korea as bastions of free-market capitalism. It is difficult to believe that the agencies and the USA would have taken such a public and hard-line approach to Pacific Asia if the Cold War were still in progress. More generally, the agencies have seized on the Pacific Asian situation as providing even more evidence of the deleterious effect of state intervention in the economy, the inefficiency of state enterprises and the need for 'good governance' in the sense of effective administrative regulation. Thus, the 1997–98 crises are presented not as a failure of the revised neoliberal interpretation of Pacific Asian development but rather as further proof of the validity of the neoliberal position and prescriptions.

The conditions attached to the 1997 and 1998 IMF 'rescue packages' for Indonesia, South Korea and Thailand require the opening of the economies concerned to international capital and competition. Restrictions on foreign ownership of companies are to be reduced or eliminated, particularly in the financial sectors. In South Korea, the IMF has insisted that foreign banks be allowed to establish subsidiaries and brokerage houses and gain access to the domestic money and corporate bond markets (Khor 1998). There is little doubt that these moves will result in substantial increases in the foreign ownership and control of some Pacific Asian economies. Some observers have seen these requirements as going well beyond the normal agency macro-economic targets and reflect American and, to some extent Japanese, influence on the IMF to ensure that their companies gain a substantial presence in the 'rescued' economies (*ibid.*).

The 1997–98 crises and the agency prescriptions have to be seen in the context of the liberalisation since the early 1980s, particularly of trade and financial regimes, that has taken place in all the Pacific Asian economies. There has been continual pressure from the Western nations and international agencies for this to happen, and liberalisation has become an issue in almost all bilateral and multilateral negotiations. Indeed, it has been suggested that the USA, in particular, helped to bring about the crises by the promotion of open-market policies that made economies vulnerable to financial havoc (Sender *et al.* 1997: 69). However, it must also be stressed that the liberalisation of trade and financial regimes facilitated the export of capital and gave the domestic economy comparatively free access to international finance. There is no doubt that these measures were sought by domestic capital interests and seen by the governments as significant developmental measures. Indeed, the rapid development of Indonesia, Malaysia and Thailand from the mid-1980s rested very heavily on the liberalisation that enabled the large-scale export of capital and the relocation of manufacturing processes from, in particular, Taiwan and South Korea, and facilitated the inflow of investment and gave domestic capital access to international funds. However, it is clear with the benefit of hindsight that liberalisation made the economies more vulnerable to international and regional events outside national control, particularly in the light of the lack of effective control over the rapidly expanding and internationalised financial sectors. There is here an interesting contrast between South Korea and Taiwan. In the former, liberalisation of trade and financial regimes, including the privatisation of commercial banks, progressed rapidly after 1981, while in Taiwan, apart from the relaxation of controls over the export of capital, liberalisation made much more limited progress.[5] Taiwan very largely avoided the recession of the early 1980s and at the time of writing (June 1998) any serious effects from the 1997–98 crises.[6]

The 1997–98 crises have highlighted the whole question of state intervention and market distortion in the third generation of Pacific Asian growth economies, Indonesia, Malaysia and Thailand. Previously, much emphasis had been placed on the far more limited role played by the state and the greater significance of TNCs and foreign investment (Castells 1991: iii; Page 1994: 624; World Bank 1993: 7). It is usually asserted that Thailand since the late 1950s has been by far the least interventionist of the Southeast Asian economies. However, until the late 1980s the direct intervention of the Thai state through subsidies, market controls and tariffs was considerable by international standards (Dixon 1998a). During the early 1980s, both Thailand and Indonesia were subject to formal and informal structural adjustment policies aimed at reducing what the

[5.] Taiwan also has different industrial and trade structures from South Korea, with a large number of comparatively small firms, a lower level of dependence on regional trade and the export of capital goods and electronics.

[6.] This does not mean that Taiwan has experienced no adverse affects from the 1997–98 crisis. The growth of exports slowed during the first quarter of 1998 for the first time since the early 1980s, and between July 1997 and April 1998 the currency depreciated by 16 per cent against the US dollar.

World Bank and the IMF regarded as the high levels of state intervention (Dixon 1995b). In the event, only in Thailand was there any significant reduction in state intervention and direction of the economy (Dixon 1996, 1998a). In Indonesia and Malaysia, the state has continued to play a major interventionist role, which has again become the subject of international agency criticism and, for Indonesia, of conditionalities during the 1997–98 crises. By contrast, the developmental problems faced by Thailand during the 1990s and highlighted by the 1997 financial collapse have been linked to the more limited role played by that state compared with its neighbours. Indeed, there have been calls by planners, business interests and observers for a more interventionist and effective state (see, for example, Ammar Siamwalla 1993; Dixon 1998a). In the general enthusiasm over Thai economic growth and liberalisation, the comments by the World Bank made during the early 1980s on the inadequate and ineffective regulation of the Thai economy were forgotten (Dixon 1998a).

It should be stressed that the debt crises that emerged in Pacific Asia during 1997–98 have principally affected the private sector. In this respect, the situation is very different from that which prevailed during the early 1980s and in most of the Third World states that have subsequently been subject to structural adjustment programmes. The IMF has been extremely critical of the Pacific Asian governments concerned for their ineffective regulation of the financial sectors. Additionally, the South Korean government has been criticised for the controls that it exercised over the *chaebols* (conglomerates), which encouraged them to extend production and credit in the face of contracting markets and declining competitiveness. Thus the hitherto generally highly successful relationship between the South Korean state and the major corporations, which the international agencies had classed as 'market-friendly', has become relabelled as 'interference' and a major factor in the present crisis (see *Far Eastern Economic Review* 26/3/98: 10–15; 30/4/98: 10–14).

The IMF's immediate prescriptions for the Pacific Asian economies demand that governments play strictly by the market and not rescue ailing companies, while at the same time expecting them to ensure the paying of private sector debts. It is this that has led some commentators to assert that the IMF has changed its role, certainly with respect to the Pacific Asian growth economies (Khor 1998).

The remarkable 1997–98 *volte-face* of the international agencies over their interpretations of the development of key Pacific Asian economies might be expected to further discredit both the agencies and the increasingly confused neoliberal position. However, in the light of the amount of discrediting that has taken place previously and its limited impact, it is difficult to see what will be achieved. For while the confidence of those opposed to neoliberalism may be bolstered, they have little input into the practicalities of development at national and international levels. In addition, virtually the entire Western media presentation of the 1997–98 crises was in terms of the agency position, this being underlined by comments on the 'end of the Pacific Asian miracle' and inept and undemocratic governments.

Thus the challenge to neoliberalism posed by three generations of

Pacific Asian growth economies appears to have suffered a setback as a result of the 1997–98 series of crises. Whether there is a rapid re-establishment of miracle growth, or a period of adjustment to a pattern of slower growth, there is little likelihood of the international agencies and principal Western donors experiencing the collective amnesia that followed the recovery from the brief crises of the early 1980s. There is no longer any reason to underplay or attempt to relabel state intervention as market-friendly activity. It is difficult to see how there can be any credible retreat from the relabelling of state activity in Pacific Asia as 'interference'. The agencies' lesson for the rest of the Third World is that the intervention that characterised the Pacific Asian economies results in unsustainable growth and merely defers painful adjustments. Thus the warning contained in *The East Asian Miracle* report that other countries should not emulate Pacific Asian state intervention is underlined.

In their concern over the 1997–98 crises, the agencies and principal Western donors appear to have lost sight of the remarkable achievements of the Pacific Asian economies since the 1950s. The view appears to be that the policies and governments have long been flawed, but this has only recently been clearly exposed by the financial crises. These revealed unsound policies and the inability of governments to adjust to changed circumstances and deal effectively with the crises, or indeed to act in order to prevent the crises in the first place. Yet much literature (including *The East Asian Miracle*) has stressed the flexibility of policy in the Pacific Asian economies. In addition, since the early 1960s, the Pacific Asian region has been remarkably free of economic crises in a period when they have been endemic in the global system. In South Korea, for example, GDP grew at an average annual rate of 8.4 per cent between 1960 and 1996, only faltering during the brief debt-related crises of the 1980s.

In criticising the agency position, it is important to stress that, behind the sustained high growth rates, there *were* signs of long-term structural problems and contradictions in the development process in some, if not all, of the Pacific Asian economies. It could be argued that favourable regional and global circumstances, reinforced since the early 1980s by liberalisation of trade and financial regimes, helped to maintain growth and diverted attention from underlying problems. Here there is a need for much detailed analysis beyond the scope of the present chapter.

Particularly in the ASEAN-3 economies, the rapid growth of GDP and manufacturing exports was beginning to reveal some significant imbalances and contradictions by the mid-1990s. In Thailand, for example, the lack of long-term investment in education and training was resulting in a shortage of skilled labour, which was inhibiting attempts to follow the Asian NIEs' route to high-tech production, a situation exacerbated by the limited investment in R&D by the private sector and poor linkages with government efforts in this direction. In addition, the speed and concentration of development in the Bangkok metropolitan region, combined with the lack of substantial investment in infrastructure, was resulting in congestion and escalating production costs. Thus Thailand was losing the comparative advantage in labour-intensive manufacturing while lacking the basis for the transition to more skill- and capital-intensive activities.

The World Bank's (1993) interpretation of the Pacific Asian experience claims that development has been remarkably evenly spread between groups and areas. Indeed, there is considerable evidence to support this view in the highly unusual cases of the Asian NIEs (*cf.* Forbes 1993; Davidson and Drakakis-Smith 1997). However, this is not necessarily the case with respect to the ASEAN-3 economies (e.g. Dragsbaek Schmidt 1997, 1998; but see discussion in Booth 1997 and the material in Watkins 1998). In Thailand, for example, while there has been a marked fall in the level of absolute poverty, there is considerable evidence for a widening of income disparities as rapid export-based development has accelerated since the mid-1980s.[7] In addition, although the national situation is very different, the rapid growth of Vietnam since the end of the 1980s has been associated with rapidly widening income differentials[8] (Drakakis-Smith and Dixon 1997: 29–33).

At the Pacific Asian regional level, it can be argued that, outside the four NIEs, the Asian miracle refers to the experience of a minority of principally urban-based people. Koppel (1997: 5) refers to the 'Other' Asia, where 80 per cent of the people live and which has not been a clear beneficiary of the economic boom. Rapid growth of GDP and dramatic increases in the manufacturing sector's share of export earnings have generally not been transformed into changes in the structure of the economy or, even less, of employment. The growth of employment in manufacturing and in such areas as financial services has to be set against the continued dominance of the agricultural sectors and the expansion of informal activities. In addition, much of the urban-based service and manufacturing employment is casual (Dixon 1997). The limited generation of employment resulting from the export booms and its generally casual and insecure nature has given rise to the label 'jobless growth' (Gonzales 1997, cited in Koppel 1997: 6).

In a number of Pacific Asian economies, much growth since the early 1990s could be described as 'artificial', based heavily on the property and stock market booms and the influx of 'hot money' facilitated by liberalisation (Dixon 1997). Yoshihara (1988: 2–3) went further, describing developments, particularly in the ASEAN-3 group, as 'ersatz capitalism', which has not resulted in the establishment of 'indigenous capitalism' and which relies heavily on imported skills and technology, and often on foreign capital and capitalists. Certainly a case could be made for Thai

---

7. According to official figures, the incidence of poverty in Thailand fell from 30.0 per cent in 1975–76 to 9.6 in 1994. Over the same period, the distribution of incomes became more unequal: the Gini coefficient increasing from 0.426 to 0.525, and the income share of the top 10 per cent of the population increased from 49.3 to 57.5, while the percentage share of the bottom 10 per cent decreased from 6.1 to 4.0 (see Pranee Tinakorn 1995; Nanak Kakwani and Medhi Krongkaew 1996; Dixon 1998a). By the early 1990s, Thailand's income distribution had become one of the most unequal in the world (Pasuk Pongpaichit and Baker 1995: 202).

8. The data on Vietnamese income distribution and poverty are sparse, unreliable and the subject of considerable debate (see Drakakis-Smith and Dixon 1997: 29–33 for a discussion). On the basis of official figures, Khanh (1994) concluded that the ratio of the incomes of poorest and richest groups increased from 3–4 during the late 1970s to 7–8 during the 1980s and in 1993 to 20 for rural areas and 40 for urban areas. These changes are concealed by the comparatively modest (0.43) level of the Gini coefficient in 1994 (Padmini 1995: 2).

economic growth since the early 1980s containing ersatz elements and, since 1993, becoming increasingly artificial (Dixon 1998a).

The IMF's proposed solution to the 1997–98 crises is further liberalisation and more regulation. However, it could be argued that this would make the economies even more vulnerable to financial and related crises (see, for example, *The Guardian* editorial on 3 April 1998). While it is difficult to deny that there have been serious deficiencies in the regulation of the financial sectors, and in some cases, decision making, it is not immediately clear how these can be remedied quickly, particularly in the context of further liberalisation. Under prevailing global conditions, the regulation of the financial sector has proved increasingly difficult and unsatisfactory at the national level, as the recent experience of many Western countries has demonstrated. More generally, the IMF prescription separates regulation at the national level from the nature, activity and effectiveness of the state.

The 1997–98 crises, the ending of the Cold War and the perception by the West that the dynamic Pacific Asian economies represented an increasing threat may well have given some new life to the discredited and confused neoliberal position. Further, the challenge posed by the ATEs may also be compromised by a slowing of their principally regionally based growth and embroilment in a Pacific Asian recession. Such a situation may well undermine their position, perhaps forcing them to follow international agency prescriptions for the reduction of state activity more closely.

Developmental implications

The most basic weaknesses in the neoliberal interpretation of the development of Pacific Asian economies concern the preoccupation with internal policy, its separation from external factors and the extremely nebulous nature of the various formulations of key concepts such as 'external orientation' and 'open economy'. These issues are all apparent from a careful reading of *The East Asian Miracle* report and been have extensively commented on by Gore (1996) in particular.

The neoliberal position denies any influence of the structure of the global economy on the economic performance of individual countries (*ibid*.: 82) and downplays the influence of internal factors other than policy. Since all countries experience the same external environment, differences in performance must reflect internal policies and the way in which the external environment is dealt with. Thus the global economy is depicted as a series of 'black boxes' floating in an increasingly borderless NLIEO (neoliberal international economic order). The understanding of the internal workings of the boxes is regarded as of little importance to either the understanding of, or prescriptions for, development. Such a conceptualisation assumes that the interaction between internal and external capital is a uniform process that is only affected by variations in internal policy. The more open a country is to the NLIEO the better (in the long run) will be its economic performance. There is new life here for that long-discredited concept 'comparative advantage'.

The concept of the NLIEO was first put forward by Lal (1980, cited in Gore 1996: 85), and its implementation has rested on the liberalisation of the trade and financial regimes of Third World countries in particular. This process has been furthered under structural adjustment programmes, other international agency loan conditionalities and a range of bilateral and multilateral negotiations, including rounds of the GATT[9] (General Agreement on Tariffs and Trade) (Bergsten 1996). Counter to these developments has been the establishment of the EU (European Union) and NAFTA (North America Free Trade Area) blocs. This suggests that the 1988 prediction by GATT that the global economy was moving in the direction of operation within a series of highly managed trading blocs is being fulfilled, certainly for the West (see Thurow 1992; and World Bank 1993: 364 for a rebuttal). However, within Pacific Asia, prospects for the further development of regional organisations, through ASEAN or APEC (Asia-Pacific Economic Co-operation) are presented as following the neoliberal international agenda rather than leading to a third managed trade bloc to stand alongside the EU and NAFTA (Bergsten 1994). Thus for Pacific Asia a path of 'open regionalism' is advocated; under this, the regional agenda (including economic integration) can be fulfilled while remaining open to extra-regional trade and investment (Bergsten 1997).

Overall, the IMF demands for the further liberalisation and opening of the Pacific Asian economies that have followed the 1997–98 crises progress the NLIEO agenda. There appears to be an assumption that, in the absence of significant barriers to trade and financial movements and with the required level of national regulation, the system becomes self-adjusting to a considerable degree. The neoliberal position is that it is the barriers and internal policies that create crises by preventing timely adjustment to changing national and international circumstances. There is here the reiteration of the Friedman and Friedman (1980) view of the contrast between the flexible, dynamic market and the rigidities resting from centralised control of the economy. The agencies' position is that flexibility increases with openness and the best way to improve a country's ability to deal with external changes is to increase exposure to them (Gore 1996: 8).

The neoliberal view of countries maximising their advantage and stability through open economies under an NLIEO also stresses the primacy of manufacturing exports. However, it is difficult to see where there would be markets if significantly more countries entered this field of activity (for a dismissal of this view see World Bank 1993: 361–362). Even if there were markets, in the changed global circumstances of the late 1990s, it is highly debatable whether the manufacturing export strategy with little or no protection of the domestic economy is desirable or even practical. Even in as advanced a manufacturing economy as the USA it has been found that the protection of such sectors as automobiles,

---

[9.] The furtherance of this appears to be the central purpose of the World Trade Organisation (WTO).

machine tools, steel, semiconductors and textiles has resulted in significant increases in productivity (Tonelson 1994). The protection of these American industries is referred to as 'import relief' and 'beating back predatory trade'. This type of policy contrasts with what is depicted as unacceptable protection of Pacific Asian markets, and those of the Third World as a whole, which excludes American products and fosters inefficiency.

For the agencies and the neoliberal position as a whole it is the *form* of Pacific Asian development, i.e. export-oriented manufacturing development, that has become central. Under the national policy-centred approach there is no place for debate over whether the Pacific Asian form was a product of the interaction of particular national, regional and international circumstances that no longer exist (this is denied by the World Bank 1993: 366). Given the general ignoring by neoliberalism (certainly until the 1997–98 crises) of the specifics of the Pacific Asian developmental routes, including the 'dark side of the miracle', there is no debate over whether it is even desirable for other countries to attempt to follow any of the Pacific Asian developmental paths.

While the revised neoliberal interpretations and prescriptions that have become current during the 1990s admit to intervention on the part of the state, this is seriously downplayed and the focus remains on the nature of the policies. Little attention is paid to the context in which the policies are implemented or the agencies that operationalise them, despite evidence that many of the policies considered to be critical to Pacific Asian success were implemented elsewhere with a conspicuous lack of success (see for example Jenkins' 1991 comparison of East Asia and Latin America). Indeed, even within *The East Asian Miracle* report, there is evidence that a number of the 'unsuccessful' economies had lower levels of price distortion than some of the Pacific Asian growth economies (World Bank 1993: 301). The critical question is why were policies effective in Pacific Asia but not, for example, in Latin America? This question cannot be answered within the neoliberal critique. Rather as Jenkins (1991: 200–201) and Kiely (1998: 79) have stressed, the real question is why have some states effectively promoted development while others, implementing similar policies, appear to have inhibited it? For neoliberalism, such a question is effectively precluded by the neoliberal formulation of the state. The effectiveness of the state is measured solely in terms of its degree of intervention in the economy (*ibid*.: 79). Thus the state exists in a near vacuum as a managerial agency separated from its social context.

The unsatisfactory nature of the neoliberal conceptualisation of the state and the related narrow policy prescription has been highlighted by the emergence of the transitional economies. Here the neoliberal view of the situation is that the visible apparatus of Western-style states and preferably, democratic process, needs to be established as rapidly as possible. The resultant problems, the ineffectiveness of many state institutions, and general economic, political and social confusion are seen as an inevitable part of the transition process and/or as the result of the slow and incomplete nature of the reform process. There is an element here of the state being expected to reform itself before it actually exists in the form that would make the process effective. This further exposes the contradictions of the neoliberal conceptualisation of the state. For on the one hand the agents of the state are depicted as motivated by self-interest and therefore little concerned with reform, while on the other hand they are expected to implement reforms that will seriously undermine their position.

Overall, it must be stressed that for any analysis of the Pacific Asian experience that is not predicated on the neoliberal position, the long-term development of the region's economies provides a very strong argument for putting the state back at the centre of the development debate. In terms of both development theory and practice, the questions of policy become secondary to the social formations of the state through which policy is formulated and implemented. Here there is an urgent need for a detailed, comparative investigation of the developmental state and a reappraisal of the whole issue of the necessity of effective state intervention for states that seek to industrialise 'late' (Dragsbaek Schmidt 1997; *Third World Quarterly* 1996; Hikino and Amsden 1994; Appelbaum and Henderson 1992; Wade 1990; White and Wade 1988; Johnson 1982; Gerschenkron 1962). While this runs counter to even revised neoliberal orthodoxy, this is in serious disarray. Indeed, Kwon (1994: 635, cited in Kiely 1998: 74) has described *The East Asian Miracle* report as:

> almost a text book example of neo-classicists visibly confused but too proud to admit their failure – having been so quick to blame government for economic failure in the past, they are now reluctant to admit a positive role for government in a successful economy.

While the 1997–98 Pacific Asian crisis may have given a boost to neoliberal development orthodoxy, the inherent contradictions and confusion of the whole position have perhaps become more acute.

References

Ammar Siamwalla (1993) The institutional and political bases of growth inducing policies in Thailand. Paper presented at the Fifth International Thai Studies Conference, SOAS, University of London.

Amsden, A.H. (1989) *Asia's Next Giant – South Korea and Industrialisation.* Oxford University Press, New York.

Amsden, A.H. (1994) Why isn't the whole world experimenting with the East Asian model of development: review of the East Asian miracle, *World Development* **22**: 627–633.

Appelbaum, R.P. and **Henderson, J.** (1992) Situating the state in the East Asian development process. In *idem* (eds) *States and Development in the Asian Pacific Rim*, Sage Publications, Newbury Park, Calif., 1–26.

Auty, R. (1990) The impact of heavy-industry growth poles on South Korean spatial structure, *Geoforum* **21**: 23–33.

Auty, R. (1992) The macro-impact of Korean HCI drive reevaluated, *Journal of Development Studies* **28**: 24–48.

Beresford, M. (1993) The political economy of dismantling the 'bureaucratic' centralism and subsidy system in Vietnam. In Hewison, K., Robison, R. and Rodan, G. (eds) *Southeast Asia in the 1990s: Authoritarism, Democracy, and Capitalism*, Allen & Unwin, St Leonards, Australia, 215–236.

Bergsten, C.F. (1994) APEC and world trade: a force for worldwide liberalisation, *Foreign Affairs* **73**: 20–26.

Bergsten, C.F. (1996) Globalizing free trade, *Foreign Affairs* **75**: 105–129.

Bergsten, C.F. (1997) Open regionalism. In Arndt, S. and Milner, C. (eds) *The World Economy: Global Trade Policy*, Blackwell, Oxford, 29–49.

Booth, A. (1997) Poverty in South East Asia: some comparative estimates. In Dixon, C. and Drakakis-Smith, D. (eds) *Uneven Development in South East Asia*, Ashgate, Aldershot, 45–74.

Castells, E., Gon, L. and **Kwok, R.Y.W.** (1990) *The Ship Kip Met Syndrome: Economic Development and Public Housing in Hong Kong and Singapore.* Pion, London.

Castells, M. (1991) Guest editor's editorial, *Regional Development Dialogue*, 12(1): i–vi.

Chen, E.K.Y. (1979) *Hyper-Growth in Asian Economies: A Comparative Study of Hong Kong, Japan, Korea, Singapore and Taiwan.* Macmillan, London.

Cummings, B. (1987) The origins and development of Northeast Asian political economy: industrial sectors, product cycles, and political consequences. In Deyo F.C. (ed.) *The Political Economy of the New Asian Industrialisation*, Cornell University Press, Ithaca, NY, 44–83.

Davidson, J. and **Drakakis-Smith, D.** (1997) The price of success: disadvantaged groups in Singapore. In Dixon, C. and Drakakis-Smith, D. (eds) *Uneven Development in South East Asia*, Ashgate, Aldershot, 75–99.

Dixon, C. (1991) *South East Asia and the World Economy.* Cambridge University Press, Cambridge.

Dixon, C. (1995a) Structural adjustment in comparative perspective: lessons from Pacific Asia. In Simon, D. *et al.* (eds) *Structurally Adjusted Africa*, Pluto, London, 202–228.

Dixon, C. (1995b) Origins, sustainability and lessons from Thailand's economic growth, *Contemporary Southeast Asia* **17**: 38–52.

Dixon, C. (1996) Thailand's rapid economic growth: causes, sustainability and lessons. In Parnwell, M. (ed.) *Uneven Development in Thailand*, Ashgate, Aldershot, 28–48.

Dixon, C. (1997) Some economic problems of sustainable urbanisation in South East Asia. Paper presented at the European Science Foundation

Workshop on Sustainable Urbanisation In South East Asia, 17–19 September, University of Liverpool.

Dixon, C. (1998a) *Thailand: Uneven Development and Internationalisation.* Routledge, London.

Dixon C. (1998b) Regional integration in South East Asia. In Grugel, J. and Hult, W. (eds) *New Regionalism and the Third World*, Routledge, London.

Dixon, C. and **Drakakis-Smith, D.** (1995) The Pacific Asian region: myth or reality? *Geografiska Annaler* **77B**(1): 75–91.

Dragsbaek Schmidt, J. (1997) The challenge from South East Asia: social forces between equity and growth. In Dixon, C. and Drakakis-Smith, D. (eds) *Uneven Development in South East Asia*, Ashgate, Aldershot, 21–44.

Dragsbaek Schmidt, J. (1998) The custodian state and social change – creating growth without welfare. In Dragsbaek Schmidt, J., Hersh, J. and Fold, N. (eds) *Social Change in Southeast Asia*, Longman, Harlow, 40–59.

Drakakis-Smith, D. and **Dixon, C.** (1997) Sustainable urbanisation in Vietnam, *Geoforum* **28**: 21–38.

Far Eastern Economic Review (1998) 26 March, 30 April.

Forbes, D. (1993) What's in it for us? Images of Pacific Asian development. In Dixon, C. and Drakakis-Smith, D. (eds) *Economic and Social Development in Pacific Asia*, Routledge, London, 43–62.

Friedman, M. and **Friedman, R.** (1980) *Free to Choose: A Personal Statement*, Harcourt Brace Jovanovich, New York.

GATT (1988) *Annual report.* Geneva.

Gerschenkron, A. (1962) *Economic Backwardness in Historical Perspective.* The Belnap Press, Cambridge, Mass.

Gonzales, E. (1997) Equity and exclusion: the impact of liberalisation on the Philippine working class. Paper presented at the Third European Conference on Philippine Studies, Aix-en-Provence, 7–29 April.

Gore, C. (1996) Methodological nationalism and East Asian Industrialisation, *European Journal of Development Research* **8**: 77–122.

Grahner, G. and **Stark, D.** (eds) (1996) *Restructuring Networks in Post-Socialist Societies.* Oxford University Press.

Henderson, J. (1986) The New International Division of Labour and American semi-conductor production in Southeast Asia. In Dixon, C., Drakakis-Smith, D. and Watts, H.D. (eds) *Multinational Corporations and the Third World*, Croom Helm, London, 91–117.

Henderson, J. (1989) *The Globalisation of Higher Technology Production: Society, Space and Semi-conductors in the Restructuring of the Modern World.* Routledge. London.

Hikino, T. and **Amsden, A.** (1994) Staying behind, stumbling back, sneaking up, soaring ahead: late industrialisation in historical perspective. In Baumol, W.J., Nelson, R.R. and Wolff, E.N. (eds) *Convergence in Productivity: Cross-Country Studies and Historical Evidence*, Oxford University Press, New York, 285–315.

Jenkins, R. (1991) The political economy of industrialisation: a comparison of Latin American and East Asian newly industrialising countries, *Economic Development and Cultural Change* **22**: 197–231.

Jenkins, R. (1992) (Re-)interpreting Brazil and South Korea. In Hewitt, T., Johnson, H. and Wield, D. (eds) *Industrialisation and Development*, Oxford University Press, Oxford, 167–198.

Johnson, C. (1982) *MITI and the Japanese Miracle: The Growth of Industrial Policy, 1925–1975*. Stanford University Press, Stanford, Calif.

Katz, S.S. (1996) Some key development issues for transitional economies – East and West. In Naya, S.F.N. and J.L.H. Tan (eds) *Asian Transitional Economies: Challenges and Prospects for Reform and Transformation*, Institute of Southeast Asian Studies, Singapore, 9–25.

Kerkvliet, B.J.T. and **Porter, D.J.** (1995) Rural Vietnam in rural Asia. In *idem* (eds) *Vietnam's Rural Transformation*, Westview, Boulder, Colo., and Institute of Southeast Asian Studies, Singapore, 65–96.

Khanh, T. (1994) Social disparity in Vietnam, *Business Times* (24–25 September): 4.

Khor, M. (1998) A poor grade for the IMF, *Far Eastern Economic Review* (15 January): 29.

Kiely, R. (1998) The World Bank and development, *Capital and Class* **64**(1): 63–88.

Koppel, B. (1997) Is Asia emerging or submerging? Perspectives on the future of the Asian Miracle, *Nordic Newsletter of Asian Studies* **4** (December): 5–10.

Krugan, P. (1994) The myth of Asia's miracle, *Foreign Affairs* **73**: 62–78.

Kwan, C.H. (1994) *Economic Interdependence in Asia-Pacific: Towards a Yen Bloc*. Routledge, London.

Kwon, J. (1994) The East Asian challenge to neo-classical orthodoxy, *World Development* **22**: 635–644.

Lal, D. (1980) A liberal international economic order: the international monetary system and economic development, *Princeton Essays in Economics and Finance*, 129.

Livingstone, I. (1997) Industrial development in Laos: new policies and new possibilities. In Than, M. and Tan, L.H. (eds) *The Challenge of Economic Transition in the 1990s*, Institute of Southeast Asian Studies, Singapore, 128–135.

Mallon, R.L. (1993) Vietnam: image and reality. In Heath, J. (ed.) *Revitalising Socialist Enterprise: A Race Against Time*, Routledge, London, 204–222.

Myat Thein and **Mya Than** (1996) Transitional economy of Myanmar: performance, issues and problems. In Naya, S.F. and Tan, J.L.H. (eds) *Asian Transitional Economies: Challenges and Prospects for Reform and Transformation*, Institute of Southeast Asian Studies, Singapore, 210–261.

Myrdal, G. (1967–68) *Asian Drama: An Inquiry into the Poverty of Nations*. Penguin, Harmondsworth.

Nanak Kakwani and **Medhi Krongkaew** (1996) Big reductions in poverty in Thailand, *Poverty Alleviation Initiatives* **6**: 7–12.

Naya, S.F. (1996) Introduction. *Asian Transitional Economies: Challenges and Prospects for Reform and Transformation*, Institute of Southeast Asian Studies, Singapore: 1–6.

Noerlund, I. (1997) Nike and labour in Vietnam, *Nordic Newsletter of Asian Studies* (October): 15–18.

Noerlund, I. (1998) The labour market in Vietnam: between state incorporation and autonomy. In Dragsbaek Schmidt, J., Hersh, J. and Fold, N. (eds) *Social Change in Southeast Asia*, Longman, Harlow, 155–182.

Padmini, R. (1995) *Report on Social Aspects, Vietnam Urban Sector Strategy Survey*. TA 2148-VIE, Asian Development Bank, Manila.

Page, J.M. (1994) The East Asian miracle: an introduction, *World Development* **22**: 615–625.

Page, J.M. and **Petri, P.A.** (1993) Productivity change and strategic growth policy in the Asian miracle. World Bank staff paper, Washington, DC.

Pasuk Pongpaichit and **Baker, C.** (1995) *Thailand's Boom*. Silkworm Books, Bangkok.

Pranee Tinakorn (1995) Industrialisation and welfare: how poverty and income distribution are affected. In Medhi Krongkaew (ed.) *Thailand's Industrialisation and its Consequences*. Macmillan, London, 218–231.

Rigg, J. (1995) Managing dependency in a reforming economy, *Contemporary Southeast Asia* **17**: 147–172.

Rigg, J. (1997a) Uneven development and the (re-)engagement of Laos. In Dixon, C. and Drakakis-Smith, D. (eds) *Uneven Development in South East Asia*, Ashgate, Aldershot, 148–165.

Rigg, J. (1997b) *Southeast Asia: The Human Landscape of Modernisation and Development*. Routledge, London.

Rodan, G. (1989) *The Political Economy of Singapore's Industrialisation*. Macmillan, London.

Rosenberger. L.R. (1997) Southeast Asia's currency crisis, *Contemporary Southeast Asia* **19**: 223–251.

Sender, H., Jayasankaran, S. and **McBeth, J.** (1997) Not a happy bunch: World Bank/IMF meeting wasn't the party expected, *Far Eastern Economic Review* (October): 69–70.

Singer, H. (1988) The World Development Report 1987 on the blessing of outward orientation: a necessary correction, *Journal of Development Studies* **24**: 125–136.

St John, R.B. (1997) End of the beginning; economic reform in Cambodia, Laos and Vietnam, *Contemporary Southeast Asia* **19**: 172–189.

Third World Quarterly (1996) theme issue on 'The developmental state?' **17**(4): 585–706.

Thurow, L. (1992) *Head to Head. The Coming Economic Battle Between Japan, Europe and America*. Morrow, New York.

Tonelson, A. (1994) Beating back predatory trade, *Foreign Affairs* **73**: 123–135.

Toye, J. (1987) *Dilemmas of Development*. Blackwell, Oxford.

Tsiang, S. and **Wu, R.** (1985) Foreign trade and investment as boosters of take off: the experience of the four Asian NICs. In Galenson, W. (ed.) *Foreign Trade and Investment*, University of Wisconsin Press, Madison, 320–343.

Vietnam Labor Watch (1997) *Nike Labor Practices in Vietnam*, Ho Chi Minh City.

Vu Tuan Anh (1995) Economic policy reform: an introductory overview. In Noerlund, I., Gates, C. and Vu Cao Dam (eds) *Vietnam in a Changing World*, Curzon Press, Richmond, Surrey, 31–70.

Wade, R. (1990) *Governing the Market: Economic Theory and the Rise of Government in East Asia*. Princeton University Press, Princeton, NJ.

Wade, R. (1996) Japan, the World Bank and the art of paradigm maintenance: the East Asian miracle in political perspective, *New Left Review* **217**: 5–36.

Watkins, K. (1998) *Economic Growth with Equity*. Oxfam, Oxford.

White, G. (ed.) (1988) *The Developmental State*. Macmillan, London.

White, G. and **Wade, R.** (1988) Developmental states and markets in East Asia: an introduction. In White, G. (ed.) *The Developmental State*, Macmillan, London, 1–29.

World Bank (1983) *World Development Report*. Oxford University Press, New York.

World Bank (1991) *World Development Report*. Oxford University Press, New York.

World Bank (1993) *The East Asian Miracle: Economic Growth and Public Policy*. Oxford University Press, New York.

World Bank (1994) *Adjustment in Africa*. World Bank, Washington, DC.

Worner, W. (1989) Economic reform and structural change in Laos. *Southeast Asian Affairs 1989*, Institute of Southeast Asian Studies, Singapore, 187–210.

Yoshihara, K. (1988) *The Rise of Ersatz Capitalism in South East Asia*. Oxford University Press, Singapore.

CHAPTER 10

Healthy cities in developing countries
A programme of multilateral assistance

Ton van Naerssen and Françoise Barten[1]

Introduction

The rapid pace of urbanisation, the spread of urban poverty and the deteriorating physical and social environment of many cities in the South has led to a growing interest in the health status of urbanites. In this context, health does not merely refer to the absence of diseases or infirmities: following the constitution of the World Health Organisation (WHO), we will define it as a state of complete physical, mental and social well-being. Health depends on a good physical environment where people are protected from chemical and biological hazards. Traditionally, development co-operation pays more attention to the provision of health-care facilities than to the improvement of other environmental determinants of health, yet these are of crucial importance. Provision of clean water in a community, for example, can have a much greater positive effect on the health of the people than an extension of its health-care facilities.

In this chapter, we focus on the close interdependence of health and the urban environment. The latter is conceived as the complex whole of the physical, economic, social and political conditions in an urban area. After providing an overview of the relationship between health and the physical components of the urban environment, we will show how their improvement depends on other dimensions of the urban environment. We will then introduce the UNDP/WHO Healthy Cities programme in developing countries (1995–1998) as an attempt to address comprehensively the problems of urban health in the South through international development co-operation. We will highlight the basic assumptions of the programme and, as a specific case study, refer to the implementation of the programme in Dar es Salaam, Tanzania. Because of its focus on the interrelations between the several dimensions of the environment in a given (urban) geographical area, Healthy Cities is a subject of particular interest for health geographers (Verhasselt 1993, 1997; Werna *et al.* 1996).

---

[1]   The authors would like to thank Sue Houston for correction of the draft.

The relationship between urban health and the physical environment

In most countries of the South, the urban population, urban poverty and the use of energy continue to grow rapidly and thus the quality of the physical environment in the cities is deteriorating steadily. This matter is attracting increasing attention, as is proved by the environmental programmes of the United Nations Centre for Human Settlements (UNCHS/Habitat) and the number of publications focusing on the subject (e.g. Hardoy *et al.* 1992).

One of the most serious problems concerns air pollution. It has been estimated that 1.4 billion people live in cities where annual averages of sulphur dioxide levels and particulate matter exceed WHO health guidelines (UNEP and WHO 1988). Major sources of urban air pollution are transport, manufacturing industry, and coal- or oil-fired power stations. The relative importance of the sources of air pollutant emissions differs between cities. In most urban areas, it is not manufacturing industry that is the major culprit, but transport. Megacities such as Manila, Bangkok, Bombay, Mexico City, São Paulo and Lagos already have overwhelming air pollution created by traffic. It is also expected that the significance of motor vehicles as a source of pollution will increase, as most of the expected growth in the number of vehicles will occur in the developing world.

In a number of megacities, the problem is aggravated because they are surrounded by mountainous areas, which restrict air circulation. The morning smog in Mexico City is a case in point. Furthermore, vehicle fleets tend to be old and poorly maintained. In Nigeria, due to declining oil revenues, the motor vehicle fleet decreased from over 500,000 in 1982 to about 280,000 in 1992; however, an estimated 80 per cent are second-hand and in poor condition, and the resultant pollution generated is higher than in the past (Olabode 1994).

Links between health problems and air pollution levels have been suggested by comparisons between the health of people in highly polluted areas within cities and those in less polluted areas. Some of these have shown a strong association between the incidence of respiratory infections and air pollutants. Studies in Latin America and the Caribbean suggest that, given the prevailing levels of air pollution and the populations exposed, over two million children suffer from chronic coughs because of urban air pollution, and that air pollution causes an extra 24,300 deaths per year in Latin America (Romieu *et al.* 1990).

Indoor air pollution is also a matter of serious concern due to the extensive use of biomass fuels for heating, cooking and lighting. Coal, kerosene, wood and liquefied petroleum gas are widely used, especially among poorer households. Besides the risks of fire in vulnerable built-up environments, these fuels burned in poorly ventilated houses have adverse health effects. Women and children are particularly affected, as has been shown by a study in Jakarta examining the links between the indoor environment and respiratory diseases in mothers and children (Surjadi 1993).

Other major problems of the physical environment concern the provision of safe water and sanitation facilities. Insufficient quantities of potable water, as well as dirty water and inadequate sanitation, cause high rates of water-related diseases, such as diarrhoea and schistosomiasis. The contamination of water supplies also poses health risks. In many urban areas, surface and groundwater sources are contaminated by residential and industrial waste: major sources of water pollution are the leakage of dumped waste and the unsanitary disposal of excreta and collected sewage.

Most rivers flowing through the cities of the South are large open drains, containing a mixture of raw sewage and untreated industrial effluents. The River Ciliwung in Jakarta, for example, is so strongly polluted by organic waste that its oxygen content has almost disappeared. In the dry season, when the water does not flow, the estuary transforms into a huge, black, stinking pool. In many Southeast Asian countries, the rise of export-oriented industries, especially textiles, causes chemical pollution of wells and rivers. The Indian Centre for Science and Environment (1993) reports that 114 cities with 50,000 or more inhabitants dump untreated sewage into the River Ganges. To this is added industrial waste from tanneries, paper and pulp mills, petrochemical and fertiliser complexes and rubber factories.

Disposal of solid waste is a third problem area. An estimated 30–50 per cent of the solid waste generated in urban areas of the South is left uncollected (WHO 1993). Rubbish heaps build up in streets and open spaces, providing excellent breeding grounds for rodents and other disease vectors. Children, who often play on wasteland near or on rubbish, are among the first affected. Solid waste represents a major reservoir of toxic metals in the urban environment. Hazardous industrial waste poses a particular problem as it is currently dumped on open landfill sites without any control or provision to prevent human exposure. The effects are felt particularly by the people who make their living from scavenging among the dumps and disposal sites.

Since many of the urban poor live in slum and squatter areas without proper facilities and near polluted rivers and waste dumps, it is not surprising that their health is considerably worse than that of their neighbours who live in the richer residential areas. A report, made within the framework of the World Bank's Urban Management Programme, and comprising an overview of fifty case studies from secondary sources, indicates strong intra-urban differences in physical environment and health (Bradley *et al.* 1991). The findings are confirmed by international research initiated by the Stockholm Environmental Institute (SEI) comparing Accra, Jakarta and São Paulo (Jacobi 1990; McGranahan 1991; Songsore and McGranahan 1993).

In Accra, for example, the overall mortality rate is 5.5 per thousand inhabitants, but it varies from 1.3 in the highest-income area to 23.3 in one of the poor neighbourhoods. Due to infectious diseases, such as malaria and diarrhoea, three times as many under 14-year-old children die in low-income areas as in the high-income areas, and in the low-income areas twice as many older people (above 44 years) die because of heart and vascular diseases compared with the high-income areas. In the case of respiratory diseases, the difference is even greater. In São Paulo, where the differences between the rich and the poor are even more pronounced than in Accra, the researchers distinguished four zones. In zone 4, comprising less than 10 per cent of the inhabitants, the average income is three times as high as in zone 1, where 44 per cent of the population live. The chance that small children (below 4 years) will die because of infectious and respiratory diseases proves to be four times greater in zone 1 than in zone 4, and the percentage of deaths due to accidents and violence among the population is twice as high in zone 1 as in zone 4.

Social insecurity, instability and alienation characterise the dark side of life in the cities, sometimes summarised in the concept of 'urban stress'. Broken families and neglected children are specific phenomena of the big city, and related problems such as prostitution, drug abuse, alcoholism and violence constitute considerable health risk factors. They are problems that occur mostly in urban poor areas.

The relationship between health and the urban environment

Urban health depends not only on the physical environment but also on the economic, social and political environments, such as the economic profile of the city, the social and demographic composition of its population, and urban policy. Moreover, these dimensions of the urban environment are strongly determined by processes at the regional, national and global scales, such as national economic policy, distribution of incomes, power relations and globalisation. Many examples of complex patterns of relationships can be given.

The economic development of a country clearly determines urban health. A study of Kingston (Jamaica) for the period 1977–1989 notices the following changes in the labour market: a shift of employment from the public sector to the private, secondary sector (male labour to wood-based industries, women to the clothing and apparel industry), an increase in the number of street traders, and an increase in unemployment among youths aged below 15 and adults aged over 55. The unemployed are not the only ones who are poor. Due to low wages – a precondition in order to compete in the global market – a new group of working urban poor has come into existence. According to the researcher, unemployment and poverty in Kingston explain the increase in violence and assault, the major cause of disability among the male population aged between 15 and 44. Disability among women is mainly due to mental problems. Here one finds a relationship between work stress, poverty and an increase in schizophrenia (Robotham 1994).

Export-oriented manufacturing industry in East Asia provides an example of the interdependence of national economic policy and urban health. As a consequence of their export drive, most governments attach high value to attracting and satisfying foreign investors. This is the reason why existing environmental regulations are often not implemented, and independent organisations of (female) workers are being opposed. In Malaysia, one of the global centres of the production of semiconductors, tens of thousands of young women are working in the electronics industry. Since the beginning of the 1970s, the establishment of a national electronics trade union has been prohibited by the national government. Around the mid-1980s, it seemed as if such a trade union would be allowed, but under pressure from (mainly US) transnational companies a favourable decision was not taken. Workers in the electronics industry often complain of sore eyes, headaches and bad working conditions, but health NGOs working on these issues are harassed, and independent research into the health problems of the workers is forbidden. The foregoing example refers to the political environment as a determinant for health. The national political environment structures health services systems and thus influences public health. China, Nicaragua and South Africa are clear examples of countries where changes in political ideology have had important consequences for the health of their populations. During the early decades of the communist regime, the principles of the Chinese health policy served as an example for many developing countries. The Chinese health-care system was highly decentralised with skilled health workers, midwives and 'barefoot doctors' working at village level and, in the urban areas, at neighbourhood level. Neighbourhood organisations propagated preventative health measures, and traditional healing practices existed alongside Western medical methods. However, the economic and social policy reforms of 1978 and thereafter eroded the health system. Economic liberalisation led to the introduction of the private sector and 'pluralism' in health care. At the beginning of the 1980s, the countryside had already lost 3.7 million village health workers (Hillier and Zheng 1994). There is no reason to assume that the quality and affordability of health care in the urban areas has been maintained. Massive migration to the cities and the return of urban poverty, as visible in the streets, suggest the opposite.

Socio-economic policy in many developing countries, especially in the poor African countries, is determined largely by international donors. The structural adjustment programmes (SAPs) of the World Bank and the IMF are notorious since they badly affect health service systems (Phillips and Verhasselt 1994). The implementation of a SAP entails cuts in the national budget, especially in subsidised social sectors such as public health. It is the intention that, to a large degree, the public health sector will be financially self-supporting, and that private initiatives will compensate for the loss of the size and quality of the public health sector. The principle that health services are something that should be paid for naturally means a burden for the poor, simply because they will find it very difficult or impossible to pay.

Since health is so dependent on other factors, the SAP philosophy also affects the health of the urban poor in other ways. The government will be less supportive of social housing programmes, education, improvement of infrastructure and food subsidies, which are all matters of crucial importance for health. The introduction of a SAP usually implies an increase in unemployment, due to redundancies in the public sector and in those parts of the private sector that are not able to cope with import restrictions and international competition. All in all, the implementation of a SAP departs from a narrowed vision on health, supposedly emphasising the quality and improvement of biomedical (curative) care but neglecting other determinants. However, quality of health care refers not only to efficiency and effectiveness but also, primarily, to equity. Participation and intersectoral co-ordination are crucial in that sense.

Urban management and policy are also of great influence on health. Today, the tendency exists to decentralise decision making and finance to the local – municipal or district – level. This can be advantageous. For example, in the old system a disproportionate part of the government budget used to go to the capital city. Giving the secondary cities more importance guarantees to a certain degree that they will get a fairer share. Decentralisation, however, is often a hidden form of saving, whereby the local level is supposed to organise its own income, including, among others things, charging fees for health services.

In some circumstances, decentralisation of urban management can weaken the trade unions in the health sector. In 1993, the introduction of the new Local Government Code in the Philippines created unrest among workers in the health sector, since the code implied that on certain issues the trade unions had to deal with local authorities instead of the central government. However, the strength of the unions at the local level varied considerably.

Participation is one of the key notions of current urban management. The better the people are organised, the more they can pressurise the local government, thus increasing the potential benefits for the poorer segments of the population in matters such as education and public health. Moreover, participation means sharing the responsibility for development projects, and in this way their sustainability will be enhanced. However, the organisational capacity of local people often depends on the democratic quality of the national state, which shapes the urban political environment in which the existence of independent non-governmental organisations and community-based organisations will be prevented or encouraged. Unfortunately, as the example of Malaysia demonstrates, in many countries trade unions and interest organisations, including health NGOs, are controlled by the national government.

Towards an integrated approach

The examples above demonstrate that any approach to solving health problems in urban areas must take the political, social and economic dimensions of the urban environment into consideration. This explains the plea for a broad, integrated approach towards health. The bottlenecks hindering an improvement in health are not so much at the health-care level but concern economic, political and institutional matters in the first place. A biomedical approach has to be replaced by one which gives full attention to all determinants of health.

In this respect, the Alma-Ata Conference of 1978, organised by the

WHO and UNICEF, meant a milestone in the history of health improvement. The conference resolution 'Health for All in 2000' and the 'Declaration on Primary Health Care' are of major importance. According to McDonald (1992), it is possible to see the framework of a new health-care model with universal significance in the broad lines of the primary health care (PHC) approach as laid out at the Alma-Ata Conference. PHC was presented by the conference as a health system integrated into the development plans and programmes of a nation:

> It forms an integral part both of the country's health system, of which it is the central function and main focus, and of the overall social and economic development of the community.

(WHO 1978: vi)

The integrated approach pays full attention to all determinants of health. It places the causes of bad health within the context of the society and considers poverty alleviation as a major strategy in the improvement of health. It follows that urban programmes in PHC involve the aggregate of activities that combines curative and preventive health interventions with programmes for urban environmental improvement. The latter include programmes concerning the physical environment as well as income-generating programmes and socio-political programmes promoting empowerment of social groups and communities.

The key notions of the integrated PHC approach are equity, prevention, intersectoral co-operation and community involvement (comprising the related concepts of participation and empowerment). All four key notions meet resistance. In the current era of neoliberalism and privatisation, policy makers are inclined to overlook the principle of equity in health in particular. Prevention seems a generally accepted idea, but in practice substantially more money is still allocated to curative services, and efforts to reach a more equitable distribution of financial resources between curative and preventive activities will often be opposed by vested interests of the medical–bureaucratic world. For the same reason, intersectoral co-operation between the health and other sectors, such as city planning and education, proves difficult.

Such resistance is often hidden, since the same terminology can be used while giving it different connotations. This is particularly the case with the term 'primary health care', since following SAP programmes, another approach, called 'selective PHC', has been developed. It is more pragmatic and selects health interventions on the basis of cost efficiency. The core question here is how to improve health within the limits imposed by a restricted budget. In practice, this means disease prevention through programmes such as vaccination campaigns, family planning, and mother and child care. Another example concerns the notion of participation, which is broadly accepted but in practice is often limited to consulting the 'target groups' of the urban poor or urban women. Real participation, however, demands a political process to involve people and to improve their institutional capacity to participate. This can be considered as a precondition for starting participatory processes at community level. Seen from this perspective, it is not surprising that the integrated

PHC approach meets resistance from privileged groups defending their interests.

The Healthy Cities movement, which started in the North about a decade ago, does not present clear solutions for the problems mentioned, but at least it can be said that it is a serious effort to reconcile different interests and to implement an integrated approach to urban health. Healthy Cities acknowledges the close interrelationship between urban health and the several dimensions of the urban environment, and it attaches great value to participatory interventions at neighbourhood level. The movement started in 1986, in the wake of the First International Conference on Promotion of Health, held in Ottawa under the auspices of the WHO. In the declaration of this conference, called the 'Ottawa Charter for Health Promotion', health promotion is defined as 'the process of enabling people to increase control over, and to improve their health'. It also states that 'Health promotion works through effective community action in setting priorities, making decisions, planning strategies and implementing them to achieve better health; at the heart of this process is the empowerment of communities, and the ownership and control of their own endeavours and destiny' (WHO 1986, quoted by Davies and Kelly 1993: 14, 56).

The key concepts of the Healthy Cities programmes are actually the same as those of the (integrated) PHC approach. The differences are historical, since the former originated in the North, while the latter has been discussed in the light of health problems in the countryside of developing countries in particular. Since Healthy Cities is a movement, and its idea spread more or less spontaneously to a large number of cities, a certain flexibility will be found in what a healthy city comprises. The WHO defines Healthy Cities as 'a public health approach that acknowledges social, economic and environmental dimensions of health' (WHO 1993: 7), or as 'a facilitating programme aimed at health promotion and institutional development both at the municipal and "grassroots" level. An HC-project might be a stand-alone project or health component of a larger development effort, but its objective is improving the health of urban dwellers through improved living conditions and better health services' (*ibid*.: 12). A short and suitable notion is perhaps to conceive of Healthy Cities as a process for the implementation the WHO 'Health for All' strategy in urban areas. Such a notion implies that the ultimate aim of Healthy Cities is to reach equity in urban health. The other features of Healthy Cities, such as health promotion, inter-sectoral co-operation and community involvement, are meant to serve this aim.

Although the Healthy Cities movement is mainly successful in the North, it also spread to cities in the South, especially in West Africa and East Asia. One of the reasons that makes Healthy Cities attractive to Southern cities is its correspondence with current concepts concerning the management of the large cities of the South (Devas and Rakodi 1993; van Oosterhout 1997). In most countries of the South, interventions to change the physical and built environments of cities are haphazard and badly co-ordinated. Especially in the large metropolises comprising several municipalities, new forms of management need to be developed. Again, urban management, encompassing public policy and physical planning, will gain in effectiveness when the development of the physical environment is embedded in socio-economic, cultural and political processes. For this reason, an intersectoral approach to the whole urban environment needs to be shaped. Furthermore, to avoid the pitfalls of top-down planning, decentralisation of responsibilities to local governments and participatory processes are becoming important tools of urban management.

While the Healthy Cities programmes correspond with this 'new urban management' approach, there are also differences. The urban management programmes of the international development co-operation agencies – UNCHS/Habitat, UNDP and the World Bank – strongly propagate urban governance. Urban governance can be defined

> as the exercise of political, economic and administrative authority to manage a nation's affairs. It is the complex mechanisms, processes, relationships and institutions through which citizens and groups articulate their interests, exercise their rights and obligations and mediate their differences. ... Effective democratic forms of governance rely on public participation, accountability and transparency.
>
> (UNDP 1997: 9)

This view on urban management is realised by the pursuit of co-operation between the three sectors of society: the public sector, the private business sector and civil society (NGOs and CBOs). In its programmes in the South, Healthy Cities does not refer to the concept of urban governance. Nevertheless, there are enough reasons to incorporate the concept. It would definitely strengthen the integration of urban health policies and urban management and development planning in the South, one of the background ideas of the UNDP/WHO Healthy Cities programme 1995–1998.

The UNDP/WHO Healthy Cities programme 1995–1998

There is a certain contradiction between the idea of Healthy Cities as a movement and the UNDP/WHO Healthy Cities programme for the South. The programme is a deliberate attempt to initiate health processes in urban areas with the help of outside donors. It is to be realised in five urban areas: Fayoum (Egypt), Cox Bazar (Bangladesh), Quetta (Pakistan), Dar es Salaam (Tanzania) and Managua (Nicaragua). They differ widely in population size, level of welfare and socio-political system, which makes it difficult to compare their experiences with the programme.

Discussions about the UNDP/WHO HC-programme started in September 1993 in Geneva with a meeting of the representatives of the Environmental Division of the WHO, the UNDP–LIFE (Local Initiative Facility for Urban Environment) programme, the international NGO-network HEC (Health and Environment in the Cities), and the Dutch Directorate-General of International Co-operation (DGIS).

They came together to discuss and to devise a joint UNDP–LIFE and WHO Healthy Cities programme for the South, supported by HEC and financed by DGIS. It was DGIS that initiated the participation of LIFE and HEC in the Healthy Cities programme. Because of their experiences with community-based programmes, they were meant to provide input to WHO regarding the involvement of NGOs and urban communities. One of the decisions taken was that the Healthy Cities programme would be carried out in urban areas where programmes of UNDP–LIFE, and if possible HEC, were already operating. For this reason, it is stated in the project proposal that the programme will be implemented by the WHO in collaboration with LIFE programme partners and staff. Thus the programme is an example of development co-operation between North and South, between international development agencies, and between (inter)governmental organisations and an NGO.

The LIFE programme is meant to support small-scale projects and programmes in order to improve the urban environment. It started in September 1992 and was operating in 60 cities of 12 countries of the South five years later. Priority areas for projects are water supply, solid and liquid waste management, air pollution, and environmental education.

Basic to the approach of LIFE is the concept of urban governance, and the promotion of 'local-to-local dialogue' between the several stakeholders in the urban development process. Each LIFE project should be the result of the input of the government, private business groups and NGOs/CBOs. Outside funding is rather limited, since the projects should, as far as possible, be based on local resources. It is the framework given by this approach within which the UNDP/WHO Healthy Cities programme operates. It legitimises the perception and expectation that the programme should contribute to urban governance and the strengthening of civil society in the five cities concerned.

The WHO, as the implementer of the programme, is following a strategy of stepwise institutional development in three phases:

1. Phase I (duration 3–6 months) concerns the establishment of a local task force, defined as a nucleus of several key individuals in the city who have leadership capabilities, the desire to improve the health conditions in the city and the ability to stimulate the participation of key actors such as NGOs, community groups, municipal agencies, and university and training institutions (WHO 1995: 14).
2. During Phase II (4–6 months), one aims at the establishment of a partnership task force (to replace the local task force), which includes all stakeholders in the HC process, and the appointment of a Healthy Cities co-ordinator with the objective of formulating a municipal health plan (MHP). An MHP must be conceived as a plan of action and a tool for promoting discussion and raising awareness. MHPs are set up to identify local problems and to set priorities for health intervention.
3. Phase III (24 months) comprises the implementation of the MHP. In practice, in all urban areas except one there was a delay in the implementation of the programme. Hence, at the end of 1997 it was decided to extend the programme by one year, continuing to 1998.

Besides institutional development, the WHO aims to implement small-scale projects in geographically defined urban areas which differ in extent, such as urban districts, neighbourhoods, schools and marketplaces. This is called the 'settings approach'. Examples are street drainage, sanitation, solid waste management and school medical services, in other words, projects similar to those on which LIFE works.

The more the method receives attention, the less emphasis is put on the matter of participation, or for that matter on civil society (see, for example, the case of Chittagong in Werna and Harpham 1996). This is not to deny that WHO-HC has a keen interest in developing a community-based (health) approach, but one gets the impression that the commitment of the local municipality and key figures is considered more crucial. In addition, the dynamics of a development co-operation programme demand quick, concrete results, which are difficult to reach in institutional development. At present, however, it is too early to give an overall evaluation of the experiences with the UNDP/WHO HC programme. In the following preliminary overview we will present Dar es Salaam as a case study. It must be underlined that the case is not representative, because of the differences in conditions between participating cities already referred to.

Dar es Salaam: the setting

Dar es Salaam is the commercial, industrial and administrative capital of Tanzania. According to the 1988 census, it had a population of about 1.4 million. Considering an annual growth rate of 7 per cent, its population today is estimated at nearly 3 million, migration being responsible for 40 per cent of the increase. There is a huge unemployment problem, due to the economic weakness of Tanzania, the effects of SAPs and the accompanying liberalisation and privatisation. At the socio-political level, the country is experiencing the transformation from a one-party to a multi-party system.

Since the SAP is important in the daily life of Tanzanians, we will devote some attention to its direct effects on the health system in Tanzania, where the SAP started in 1981 after strong international pressures. The share of the health sector in the national budget decreased from 7.2 per cent in 1977/78 to 4.6 per cent in 1989/90, and expenditure per capita decreased by more than one-third between 1980 and 1986. The devaluation of the Tanzanian shilling in 1986 had the effect that within a few years medicines became three to four times more expensive. In 1993, the World Bank calculated that the ratio of population to doctors increased from 19,000 in 1981 to 24,900 in 1990, and the ratio of population to health workers from 3,300 in 1970 to 5,500 in 1990. The low wages paid to health workers and doctors have led to political tensions between these professionals and the state, and to a neglect of duties by which the quality of the health services declined further. Health services, both public and private, are expensive. Research by the Tanzania Gender Networking Programme, an independent NGO, showed that on average the cost of treatment of one case of malaria is equivalent to half of a monthly salary (Lugalla 1995).

As for the urban physical environment, the following data speak for themselves. Water is an important health problem. There is a structural water shortage, culminating in a real water crisis during 1997. Only 40 per cent of the population has a piped water supply, and the water pipe system is old and dilapidated, with frequent interruptions in supply. Only 5 per cent of households are connected to a sewer, the majority of the population using pit latrines for disposal of human waste. Around 2,000 tons of solid domestic waste has to be collected each day. In fact, refuse collection rarely takes place. To give an example, at the end of 1997 only four of the fleet of fifteen trucks were available per day, so refuse collection is one of today's major urban management problems. Even at the site of Tanzania's main hospital, there is haphazard dumping of refuse. Safe disposal of hazardous industrial wastes, such as chemical liquids and fumes, does not occur.

Who are the institutionally organised stakeholders in Dar es Salaam that should be involved in the HC-programme? The following list of major projects and actors is far from complete (see Mtasiwa 1997) but nevertheless gives an impression of the complexity of the stakeholders' approach.

1. The health sector:
- Dar es Salaam City Commission Medical Office of Health;
- Dar es Salaam Urban Health Project (DUHP);
- Malaria Control Project;
- Ministry of Health, including the Directorates of Prevention and Health Planning;
- Muhimbili University Medical Faculty;
- Dar es Salaam Urban Health Group (NGO research group, member of HEC).

2. The physical planning sector:
- Dar es Salaam City Commission Department of Planning;
- Sustainable Dar es Salaam Project (SDP);
- Ministry of Land, Housing and Urban Development;
- Urban Sector Engineering Project (USEP) (a World Bank endeavour);
- Hanna Nassif Squatter Area Project (an ILO/SDP-funded project);
- ARDHI Physical Planning and Housing University.

3. Other sectors:
- UNDP–LIFE programme;
- Tanzanian Association of Non-Governmental Organisations (TANGO) and another umbrella organisation, TAGOSODE;
- International NGOs: Plan International, Red Cross, etc.;
- private sector: Rotary Club, Lions Club.

One key actor is the Dar es Salaam local government. The current City Commission resorts directly under the prime minister's office. It replaced the Dar es Salaam City Council in June 1996, when the national government dissolved the council and took control of the metropolitan area of Dar es Salaam, causing a period of uncertainty in actual decision making. The City Commission has gained a good reputation, although some of its measures, for example concerning the 'cleaning' of the city by driving out street traders, are controversial.

Two other major actors are the Dar es Salaam Urban Health Project (DUHP) and the Sustainable Dar es Salaam Programme (SDP). The DUHP is funded by the Swiss Development Corporation, with the Swiss Tropical Institute as implementer. Its original aim was to rehabilitate the health facilities in Dar es Salaam, but in due course the scope of the project has been widened, and the project is moving from the improvement of health-care services towards a more preventative and community-based approach to urban health. As such, it is beyond doubt that the DUHP provides good opportunities for the support of a Healthy Cities project.

The SDP was the first project that started under the responsibility of the long-term Urban Management Programme of UNCHS (Habitat), UNDP and the World Bank. It offers an institutional framework within which to approach urban environmental priority areas, such as solid waste, coastal protection and squatter settlements. In 1995–1996, thirty working groups existed, involving more than a hundred people from twelve departments of Dar es Salaam national government ministries, international agencies and so on. The SDP managed to raise several million US dollars from various sources for its projects. However, the transformation of an internationally supported programme to a locally sustained one proved to be a very difficult process. In the course of 1997, the SDP had lost much of its appeal.

Mention should also be made of the NGO/CBO sector in Dar es Salaam. Actually this sector is still in its build-up phase, although much progress has been made. In 1986, the existing NGOs were mainly associations for disabled people, and charity clubs such as Rotary and the Lions, while in 1994 a directory of TANGO mentions, among others, forty-two environmental NGOs, thirty-six women's NGOs and twenty-two health NGOs in Tanzania. At the neighbourhood level, the community-organising system inherited from the days of the one-party system is still in existence, which can facilitate communication at the lower levels. Moreover, there are many informal, 'invisible', initiatives at the local level (Cranenburg 1995: 27).

Preliminary remarks on the Dar es Salaam Healthy Cities programme

Halfway through the HC programme period (July 1997) the major aim of Phase I had not been reached. The setting up of a local task force co-ordinating the programme at the metropolitan level proved to be difficult. Besides the problem of the shift in local government authority mentioned earlier, it took time to get major stakeholders together. This is one of the disadvantages of starting a new initiative in a capital city, where many institutions must be involved, and where a national government has a clear interest in what happens. Here, a third reason can be added. There was much uncertainty about the relationship between the HC programme and the SDP. Nearly all the SDP priority areas are relevant and open to a Healthy Cities settings approach, which suggests that health policies and health promotion in Dar es Salaam should take place in close co-operation with the urban management project. The municipal health plan, in particular, should be part of an all-encompassing environmental plan for Dar es Salaam. However, the parties involved had different interests, and establishing this kind of intersectoral co-operation could not easily be achieved.

Actually, intersectoral co-operation is proving difficult in nearly all Healthy Cities projects. One reason is that health institutions often consider themselves to be the prime movers of Healthy Cities programmes. City planning should have a central position in Healthy Cities initiatives, but professional traditions and vested interests in many cities are preventing Healthy Cities co-ordination from taking place within city planning departments.

After a few years, a 'consultative workshop to formulate a city health plan under the Dar es Salaam Healthy Cities project' was held in Dar es Salaam in September 1997. The workshop was organised by the Dar es Salaam City Commission in collaboration with the WHO and the Ministry of Health. It received high-level support direct from the prime minister's office, and members of the Dar es Salaam City Commission took an active part in the discussions. Papers were presented on water supply, solid waste collection, housing, occupational health, nutrition and mental health. Over three days, around sixty participants discussed the presentations, and action plans were being devised in eight working groups. In this way, the basis was laid for a municipal health plan. At the project level, four settings for intervention were selected: food markets, schools, small industrial areas and unplanned settlements.

A joint evaluation of UNDP–LIFE and WHO, shortly after the consultative workshop, valued the commitment of national and local government highly. However, there were questions about the participation of NGOs and CBOs. As a matter of fact, only a few representatives of the private business sector and NGOs/CBOs were present at the consultative workshop. It is difficult to avoid the impression that neither the WHO nor the governmental institutions in Tanzania had a clear overview of health-related NGOs and health initiatives at the community level in Dar es Salaam. It is also striking that the question of equity in health was not on the agenda, and that no mention was made of a recent study on socio-economic, environmental and health differences between the three districts of Dar es Salaam. These matters reveal that one of the basic problems in Dar es Salaam is to strengthen the NGO and CBO sector in order to create a situation in which they will not be overlooked or can forward their demands. Only participatory processes from the bottom up can realise sustainable municipal health plans and healthy urban 'settings'.

Conclusions

We have argued that planning the provision of health services alone is not sufficient to cope with pressing health problems in the cities of the South. Instead, an integrated approach to improving health is urgently needed. The Declaration on Primary Health Care (Alma-Ata 1978), the

Ottawa Charter of Promotion of Health (Ottawa 1986) and the Healthy Cities movement are important steps in the direction of such an approach. They also fit into the paradigm of urban governance that is currently in use in the 'new' urban management. In order to reach effective and sustainable interventions for improvement of the urban environment and health, it is necessary to bring Healthy Cities together with urban management programmes. The UNDP/WHO Healthy Cities programme can be considered as such an endeavour but, as the programme in Dar es Salaam shows, for various reasons it is proving to be a difficult task. One of the basic difficulties is that an integrated approach presupposes a strong civil society with a diversity of independent NGOs and CBOs. Where such a civil society is lacking it could well be that the Healthy Cities programme should aim at supporting the creation and the strengthening of new (health) NGOs and community organisations. Working at the institutional level is important and, for health geographers, a challenging field for research and applied geography, too.

References

Barten, F. and **van Naerssen T.** (1995) Towards an urban health paradigm. In van der Velden K. *et al.* (eds) *Health Matters. Public Health in North–South Perspective,* Amsterdam, Koninklijk Instituut voor de Tropen, 129–141.

Bradley, D., Cairncross, S. *et al.* (1991) *A Review of Environmental Health Impacts in Developing Country Cities.* Urban Management Programme Paper No. 6, World Bank, Washington, DC.

Centre for Science and Environment (1993) *The State of India's Environment: A Citizen's Report.* New Delhi.

Devas, N. and **Rakodi, C.** (1993) *Managing Fast Growing Cities. New Approaches to Urban Planning and Management in the Developing World.* Longman, Harlow.

Davies, J.K. and **Kelly, M.P.** (1993) *Healthy Cities: Research & Practice.* Routledge, London.

Hardoy, J.E. and **Satterthwaite, D.** (1989) *Squatter Citizen: Life in the Urban Third World.* Earthscan, London.

Hardoy, J.E., Mitlin, D. and **Satterthwaite, D.** (1992) *Environmental Problems in Third World Cities.* Earthscan, London.

Hillier, S. and **Zheng, X.** (1994) Rural health care in China: past, present and future. In Dwyer, D. (ed.) *China. The Next Decades,* Longman, Harlow, 95–115.

Jacobi, P. (1990) Habitat and health in the municipality of São Paulo, *Environment and Urbanization* **2**(2): 33–45.

Lugalla, J.L.P. (1995) The impact of structural adjustment policies on women's and children's health in Tanzania, *Review of African Political Economy* **63**: 43–53.

McDonald, J. (1992) *Primary Health Care: Medicine in its Place.* Earthscan, London.

McGranahan, G. (1991) *Environmental Problems and the Urban Households in Third World Cities.* Stockholm Environment Institute, Stockholm.

Mtasiwa, D. (1997) Who is doing what in the area of health in Dar es Salaam. Paper presented at the Consultative Meeting on the City Health Plan, 3–5 September, Dar es Salaam.

Mtani, A. and **Daffa, J.M.** (1997) An overview of environmental profile for Dar es Salaam City, 1977. Paper presented at the Consultative Meeting on the City Health Plan, 3–5 September, Dar es Salaam.

Olabode O.A. (1994) Transportation and health in the context of sustainable development: the Nigerian case. In Folasade Iyun, B. *et. al.* (eds) *The Health of Nations: Medicine, Disease and Development in the Third World,* Aldershot, 110–120.

Phillips, D.R. and **Verhasselt, T.** (1994) Introduction: health and development. In Phillips, D.R. and Verhasselt, Y. (eds) *Health and Development,* Routledge, London, 3–32.

Robotham, D. (1996) Redefining urban health policy in a developing country: the Jamaica case. In Atkinson, S., Songsore, J. and Werna, E. (eds) *Urban Health Research in Developing Countries: Implications for Policy.* CAB International, Oxford, 31–42.

Romieu, I., Weitzenfeld, H. and **Finkelman, J.** (1990) Urban air pollution in Latin America and the Caribbean: health perspectives, *World Health Statistics Quarterly* **23**(2): 153–167.

Songsore, J. and **McGranahan, G.** (1993) Environment, wealth and health: towards an analysis of intra-urban differentials within Greater Accra Metropolitan Area, Ghana, *Environment and Urbanization* **5**(2):10–24.

Surjadi, C. (1993) Respiratory diseases of mothers and children and environmental factors among households in Jakarta, *Environment and Urbanization* **5**(2): 78–86.

Surjadi C., McGranahan G. *et al.* (1994) *Household Environment Problems in Jakarta.* Stockholm Environment Institute, Stockholm.

United Nations Development Programme (UNDP) (1997) *Reconceptualising Governance.* Discussion Paper 2, Management Development and Governance Division, Bureau for Policy and Programme Support, New York.

United Nations Environment Programme (UNEP) and **World Health Organisation** (WHO) (1988) *Assessment of Urban Air Quality*. Nairobi/ Geneva.

van Cranenburg, O. and **Sasse, R.** (1995) *Tanzania. NGO Country Profile 1995*. GOM (Gemeenschappelijk Overleg Medefinanciering), Oegst-geest.

van Naerssen, T. (1995) Report of a visit to Dar es Salaam April 4–20. Report submitted to World Health Organisation, Geneva.

van Naerssen, T. (1996) Report of the second visit on behalf of WHO Healthy City Project to Dar es Salaam, 13 February – 3 March, 1996. Report submitted to World Health Organisation, Geneva.

van Oosterhout, F. (1997) The challenge for urban management in developing countries. In van Naerssen, T. *et al.* (eds) *The Diversity of Development: Essays in Honour of Jan Kleinpenning*. Van Gorcum, Assen, 221–227.

Verhasselt, Y. (1993) Geography of health: some trends and perspectives, *Social Science and Medicine* **36**(2): 119–123.

Verhasselt, Y. (1997) Geography of health in developing countries. In van Naerssen *et al.* (eds) *The Diversity of Development; Essays in Honour of Jan Kleinpenning*. Van Gorcum, Assen, 241–246.

Werna, E. and **Harpham, T.** (1996) The implementation of the Healthy Cities project in developing countries: lessons from Chittagong, *Habitat International* **20**(2): 221–228.

Werna, E., Blue, I. and **Harpham, T.** (1996) The changing agenda for urban health. In Cohen M.A., Ruble B.A. *et al.* (eds) *Preparing for the Future: Global Pressures and Local Forces*, Johns Hopkins University Press, Baltimore, 200–201.

World Health Organisation (WHO) (1978) *International Conference on Primary Health Care*. Geneva.

World Health Organisation (WHO) (1993) *The Urban Health Crisis: Strategies for Health for All in the Face of Rapid Urbanization*. Geneva.

World Health Organisation (WHO) (1995) *Building a Healthy City: A Practitioner's Guide*. Prepared by Unit of Environmental Health, Office of Operational Support, Geneva.

CHAPTER 11

Popular organisation, local power and development

María Verónica Bastías Gonzalez

Introduction

This chapter introduces the experience of co-operation for development in South America of Norwegian People's Aid (NPA), a Nordic NGO founded in 1939 by Norway's trade unions for the purpose of supporting reconstruction efforts after the Second World War from a working-class perspective. NPA's humanitarian principles are deeply rooted in the struggle against injustice and the defence of needy people. Currently, NPA works in thirty-three countries worldwide, delivering humanitarian aid and development assistance to meet the requirements of any given situation. In Latin America, it has three local representations: Guatemala, Nicaragua and Chile; and each office works in various countries according to its respective area of influence.

NPA's strategic position is set out in its Basic Guidelines papers, which provide a sound analysis of the global political and economic situation. On this basis, it assumes the institutional position as a critical actor in the debate over the role played by international co-operation. NPA contributes to overcoming major developmental obstacles, such as poverty and environmental degradation, resulting from both an unfair international economic order and the hardships of current democratisation processes in the South. It is not easy to implement this position, since it must be translated into concrete action, taking account first of all the aspirations and actual expectations of Southern actors. In this struggle, consideration must be given to their capacities, as well as the limitations of an organisation that strives to be consistent with its principles and the mandate of its founders and main partners – Norway's trade unions – while also bearing in mind Chile's domestic problems.

Co-operation for development is a long-term process that is frequently hard to reconcile with the demands of short-term emergencies, fuelled by an increasing number of armed conflicts and natural disasters around the world. If we examined this global picture carefully, we would realise that such calamities spring from the present world order, as a result of power struggles, a mindless exploitation of resources and the unfair distribution of goods in the same international economic order.

Since NPA is a small actor and has its own limitations, it has realised the need to make permanent efforts, together with its counterparts, to learn from the common experiences in a quest for new alternatives to face the big challenges posed to us by co-operation for development. What follows summarises some thoughts about our experiences in Chile and Ecuador. The Chilean representation was established in 1990, taking over the programmes that were already in place in Ecuador, Bolivia and Chile. Consistent with values such as unity, solidarity, human dignity, peace and freedom, both the method and basic approach of the organisation is to be assistance for self-reliance. This policy is outlined in Norwegian People's Aid (1993, 1994).

A brief historical review

While refraining from elaborating in too much detail, a brief historical overview will help to explain the various development processes currently under way in Latin America. We cannot overlook the impact of the Cuban Revolution, which played a singular role in influencing the development of Leftist progressive movements and proposals. Meanwhile, a labour movement, which grew thanks to a new relationship of exchange terms generated by the import substitution model implemented in the 1950s in several Latin American countries, picked up strength.

Perceiving these developments as a threat, international economic groups joined national power groups in the 1970s to encourage an increasing series of military interventions in the nation-states (which in some countries occurred later than in others) in an effort to suppress the progressive social base. The 1980s, which found some countries back on the democratic track, went into the history books stigmatised as the 'lost decade', due to a huge increase in the external debt throughout the region, the suppression of subsidies and incentives, and decline in public expenditure, among other things. Although most of the loans had been taken out by the private sector, the external debt crisis led to the implementation, throughout the region, of structural adjustment programmes (SAPs), dictated by the World Bank and the International Monetary Fund as a condition for the national governments of the region, and the detrimental effects of which hit mainly the poorest people in Latin America (*cf.* Chapter 9).

Thus, even earlier than in other poor regions of the world, Latin America was gradually driven into the process of so-called economic globalisation, by means of adjustment plans and measures aimed at reducing the regulatory role played by the state; promoting a modernisation process based on the privatisation of key economic areas; suppressing industrial development processes that sought to increase the added value of exports, and emphasising instead the overexploitation of natural resources; and laying the foundations of individualistic societies inspired in the competitive and culturally alienated logic of the market.

Chile was the first country in the region where economic adjustment policies were introduced on a draconian basis under the dictatorship, in order to rein in any possible expression of social dissatisfaction.

The impact of globalisation on Latin American societies

There are now more poor people in Latin America and the Caribbean than ever before: 210 million people out of a total population of 450 million live in poverty (CEPAL 1997), while the distribution of income is stagnant or regressive throughout the region. Economic and institutional reforms by means of SAPs (trade opening, privatisation and market deregulation), which involved social costs, were settled by the hypothesis of productive reconversion. This, it is assumed, will solve the increasing short- and medium-term unemployment rates through the economic dynamism produced by the growth of macro-economic indicators. This is the basis of the trickle-down effects of the model.

The latest report issued by the United Nations Economic Commission for Latin America and the Caribbean (CEPAL) shows that although the majority of the countries in the region have attained annual growth rates of between 3 and 4 per cent in the 1995–1997 period, progress made in the struggle against poverty has been limited or actually regressive. Latin America thus lags behind all other regions of the world. Hence, far from generating more equity, policies and reforms aimed at overcoming the ravages of the 'lost decade' had a negative effect in seeking to equilibrate the income profile distribution in the different countries of the region (CEPAL 1998).

Neoliberal policies, which have a purely economistic perspective, continue to gain ground in Latin American countries. The social cost has fostered the fragmentation of civil society, the overexploitation of natural resources, deregulation and the suppression of labour safeguards, the privatisation of public services, and an unsafe environment, among others, producing a high concentration of wealth, increased environmental degradation and erosion, and increasing poverty, which is sometimes disguised by means of manipulated social indicators.

The Chilean experience is particularly relevant in this respect, since sustained economic growth, which started over ten years ago during the Pinochet dictatorship, has turned it into a case to be replicated in a model widely disseminated among the poorest countries in the region. While being invited to imitate it, these countries were told nothing about its social consequences. Meanwhile, the issue of poverty stopped being a real concern in Chile following the dismantling of the National Commission to Overcome Poverty. This step was taken in spite of an increasingly unfair distribution of national wealth promoted by prevailing free-market policies. The latest survey on the state of the nation in the social and economic fields (*CASEN* 1998) shows Chile as the third worst-ranked Latin American country in terms of income distribution (the richest quintile earns 14.6 times more than the poorest quintile).

Ecuador is considered one of the least successful countries in the region, with economic growth rates under 3 per cent annually, negative employment and wage trends, high inflation, and low social expenditures, along with a deep crisis of governability and a poverty rate close to 60 per cent, according to local researchers. The majority of the Ecuadorian poor are indigenous people.

Human rights, democracy and development

In the 1970s, influences such as those emanating from Paulo Freire (1972), encouraging development staff to work with popular organisations in the areas of informal education and liberation theology, among others, fostered a participatory educational style in the context of activities carried out with grassroots organisations. After going through a series of changes, these later disseminated their seeds in order to strengthen action and decision-making capacities within social groups that had been historically discriminated against and excluded. In this fashion, tools were provided to the poorest members of society to defend themselves from exclusion, by making them aware of their rights and training them as valid inter-locutors of public and private decision-making groups.

During and after the period of military rule that engulfed most of Latin America, repression and fragmentation dominated the popular world, propelling these groups into a new confrontation with the official world. The quest to redemocratise the various national spaces and restore and protect fundamental human rights became necessary and urgent objectives. The right to life became important in the face of repression and the demand for truth and justice intertwined with the restoration of citizens' rights and democratic participation. However, the advent of democratic regimes in Latin America did not alter prevailing internal socio-economic systems, most of which were already underway ac-cording to the dictates of SAPs, whose impacts were discussed earlier.

Currently, there is no room in Latin America for a genuine democratic exercise that includes the aspirations of the popular sectors. Local power groups hold the notion that democracy is founded on formal universal electoral participation. From this point of view, genuine participation and influence are the province of those who control economic power. Official *governability* is justified by the notion that some benefits would ultimately reach all the victims of exclusion, who constitute the *unfeasible* sectors of the system.

At present, there are more certainties than doubts with respect to the consequences of the implementation of SAPs and their negative impacts on the poorest populations of the region. There is also no doubt that fundamental rights are yet to be fully enforced in the present imperfect, newly restored democracies, in particular with respect to the quest for truth and justice. But it is also imperative to underline that citizenship, to be exercised in a really genuine fashion, requires the enforcement of economic, social and cultural rights. Likewise, it is necessary to bring forward the comprehensive perspective enshrined in the right to development, as well as to postulate equity in economic distribution, so as to enable nations to recognise themselves as national communities. In 1986, the United Nations adopted the International Declaration on the Right to Development, and in spite of conceptual differences in this respect regarding the countries' positions, the UN has also highlighted the importance of the notion of sustainable human development.

Following the restoration of democracy, any comprehensive effort to improve living conditions in the popular sectors requires the creation and restoration of spaces devoted to dialogue and the *exercise of democracy*. Therefore, efforts in the area of organisational strengthening should not be aimed solely at the generation of self-management capacities, but also naturally demand to turn the grassroots organisations in question into *social actors*. Latin American democracies must be improved. They should be participatory democracies, inclusive of the aspirations of the popular groups, and should devise ways and means of achieving a fair distribution of the fruits of economic growth for the community as a whole. Democracy should foster integration when the process of growth is positive, and should have in place as a justice mechanism the appropriate way to cover any social costs whenever the growth process turns negative. Nevertheless, civil society will always be contested, as McIlwaine (1998) demonstrates with respect to El Salvador.

Co-operation for development: organisational strengthening and local power

Some NPA experiences in Chile and Ecuador

NPA co-operation began in both countries in the early 1980s. While the conditions that fostered the inception of the efforts of NPA initially differed, they converged over time around objectives such as achieving influence in local decision-making institutions, as well as generating a different distribution of power and resources at the social micro-level.

In Chile, this was done by supporting community groups organised by unemployed coal miners in an area traditionally affected by socio-economic marginality. Following the 1973 military takeover, this area was the target of harsh repressive actions, since the dictatorship considered it a red zone on account of its strong background of trade union activism. The project of Coronel and Loita in the so-called Coal Zone (located 500 km south of Santiago de Chile), which was born to meet a basic need (food subsistence), embraced other complementary initiatives. Throughout its expansion from helping to achieve food security, providing training and addressing the issue of productive conversion, this project offered a space for dialogue to an extremely marginalised population without much hope. At that time, Coronel Commune had around 90,000 inhabitants, over 60 per cent of whom were living in poverty.

This was the first NPA project following the restoration of democracy in Chile, and NPA decided to enter its leader in the race for the office of mayor in 1992. Thus, with little or no experience but great energy, it plunged into an electoral system that almost defeated it. It gained one-third of the popular vote but achieved its goal – not without some diffi-culties – when a majority of the members of the municipal council chose the new mayor.

Once in office, the mayor faced a huge information deficit and a great number of long-standing unfulfilled aspirations on the part of NPA's counterpart and supporters in the local community. The task at hand looked difficult. On one hand, there was a municipal administration that had no budget, along with a set of rules and a staff inherited from the dic-tatorship, and on the other hand there was an energetic mayor who had no tools with which to start drafting strategies in an area sorely lacking in technically trained human resources.

In Ecuador, the first counterpart that decided to field a candidate at the local level was UNOPAC, a peasant organisation in the district of Ayora, in Cayambe, that embraces more than 900 families in fifteen predomi-nantly indigenous and mostly poor rural communities. Links with NPA were established following an earthquake that demolished almost all the homes of local peasants in 1986. It ran in the local elections, hoping to win a seat on the municipal council in 1994, but it failed. In 1996, the national indigenous movement decided to target local administrations and suc-ceeded in having ten mayors, thirty-five municipal council members and eleven provincial councillors elected. UNOPAC had one councillor elected. This experience ran into some of the same problems already observed in Chile's coal zone: no knowledge about the systems and management of local administrations and a lack of technical qualifications among newly elected officers.

Very many lessons were learned, both from these processes and from the partnership between NPA and its local counterparts. Then other Chilean social organisations decided to vie for power at the local level in 1996; NPA and its counterparts were in a position to share experiences and provide tools for evaluating the meaning of that option. Coronel ran its mayor for re-election, and this time he obtained the third-highest national majority, with a 70 per cent local vote.

The current agenda of NPA in Chile and Ecuador includes several projects that raise the issue of local power as a means of gaining influence and effecting positive changes, at least in their more immediate context, since structural changes are beyond their reach, and the pervasive feeling is that power groups operating within the traditional parties are hardly willing to entertain the aspirations and demands of the poorest segments of society. But in spite of this boom, we believe that we must reflect longer and learn more lessons about the possibilities, results and roles of co-operation in this partnership set up to achieve local development by gaining power within municipalities and other regional bodies.

Social organisations, local democratisation and development

The popular sectors are currently experiencing a new phase. They do not accept being losers or non-viable social actors. Instead, they are building up their own alternatives in the periphery of the system or seeking to become interlocutors in social micro-scenarios. They use these spaces to apply pressure and to enhance their own alternatives, growing as viable social actors and influencing real redemocratisation, as well as balanced and comprehensive processes of local development capable of improving their living standards.[1] However, this democratisation exercise requires social organisations to develop a wide range of new capacities, from the organisational field to technical areas, and from the daily business to strategic affairs. Without them, the empowerment of organisations might become a double-edged sword that, far from serving the interests of the popular sectors, would neutralise their potentialities by legitimising development schemes that would serve the advantage of the traditional power groups. This issue not only requires popular organisations to make an accurate reading of their actual capacities to perform on the local level but also requires from international co-operation a flexible strategy to deal with the demands of the new challenges posed by this subject.

In one way or another, the strategies of international co-operation have incorporated popular education methodologies in their quest for alternatives through which to improve and develop the conditions of destitute social groups throughout the region. They are resorting to participatory methodologies, and initially pursuing the creation and strengthening of self-management capacities in the popular sectors. In order to ensure the sustainability of the initiatives in place, social interventions have gradually been intertwining with and enhancing the ability of historical and/or solidarity networks to face a social, economic and political context that is detrimental to marginalised groups.

We believe that NPA has had particularly interesting experiences in Chile and Ecuador. In Chile, the coal zone played a pioneering role by being the first to gain access to local power, thus laying the foundations for new projects along the same lines. Yet the issue of local power is still a new one. Therefore, we should share and widen the range of our reflections in order to enrich and objectivise the future perspectives of Latin American popular organisations in this area – fully respecting their own pace and demands – as well as to define the role of international co-operation in these processes, and particularly the role of NPA.

On the other hand, we move cautiously every time that a given subject becomes fashionable. We feel that it is worthwhile to ask ourselves whether such subjects emerge and consolidate as natural tendencies sprouting from social processes, or whether they are manipulated in order to co-opt development alternatives generated by grassroots organisations. The issue of *local* power is becoming increasingly recurrent in social projects in Latin America, where an actual tendency can be observed among social organisations, on different levels and in different countries, to put pressure on and/or join in decision-making spaces in local contexts.

It is impossible to isolate the foundations and consequences of the

---

[1] This provides an excellent illustration of the more general argument about postmodernism and social movements by David Simon (1998 and Chapter 2, this volume).

economic model from the role of those social organisations that achieve empowerment and/or decide to vie for a position in local governments. On the other hand, it would be incorrect to think that the confrontation with the model would not require the redefinition of the organisational spaces of the popular sectors, at least to a certain extent. Thus, the labour movement in Latin America, for example, has seen a dramatic reduction of its reach and influence as a result of the legal framework devised to regulate labour relations. Territorial organisations that are circumscribed within spatial limits might become totally isolated unless they join social networks. But the notion that the only available path is to gain access to local administrations is also untrue. Other positive experiences tell us that, particularly in sectoral groups (e.g. artisanal fishermen), a greater capacity to influence local policies might exist through means that are independent of local administrations.

Strategies of co-operation

In 1995, while working on our own strategy, we outlined some basic premises for co-operation for development in the framework of the local power option. We started by examining our experience, since this subject was emerging as a major challenge to be dealt with in our strategy. But there were many doubts as to the real effectiveness of any influence that may be gained, in terms of effecting favourable changes to the advantage of the needy.

> Following the perspectives set out by our counterparts and on the basis of the experience gained by NPA, we now see more chances of contributing to the development of marginalised sectors by means of a strategy aimed at *strengthening the decision-making and participatory capacities (empowerment)* of these same sectors on a local level, rather than contributing to a strategy focused on economic development per se.
>
> (NPA 1995)

The strategic objective for co-operation between NPA and South America

> To support the promotion and generation of the capacities of the popular sectors to exercise influence and make decisions (*empowerment*), by strengthening their social organisations and through their insertion and participation in local governments.
>
> (*ibid.*)

This objective would have to take into account the degree of organisational strength achieved by the bodies, as well as the development of capacities to influence and exercise power according to their own dynamics and the characteristics of the different contexts in which they operate. In this fashion, we can identify two levels: the first one has to do with the degree of strength and internal cohesion of the social organisation.

254

Generally, the organisation seeks to meet highly concrete requirements, including basic needs that initially are not necessarily related to the capacities and potentialities of the group, such as food, health, basic education and improved earnings, among others. Thus, co-operation plays an important role as a provider of feedback. The organisations must know their rights and become acquainted with the public and private institutions that are supposed to support them.

- *First level (survival strategy)*: The development of decision-making capacities.

  Depending on the immediate needs and the existing degree of organisation, strengthening social organisations implies, in some cases, two pedagogic simultaneous processes. On one hand, of relating the satisfaction of the basic needs of the target groups with the strengthening of their organisation, and on the other the process of shaping and training a social actor.

  (ibid.)

  Specific objectives for the first level:

  - Organisational strengthening to meet the basic needs of the target groups (health, fighting illiteracy, production, etc.).
  - Shaping a social actor (self-confidence, self-awareness and identity as a social and political actor).

- *Second level (strategy for development):* Influence, participation and insertion in local power institutions.

  As was mentioned earlier, not every group chooses to gain access to local governments, nor do we think that the potential for exercising influence would diminish if the organisations were cohesive and had their own agendas. Among cases of local empowerment, some counterparts manage to exercise influence on decision-making bodies, and yet do not belong to them. As a matter of fact, the chances are that by choosing to participate in a local administration, they may be exposed to viciously corrupt actions and electoral hanky-panky unless they are strong enough and soundly prepared from a technical point of view. Specific objectives for the second level:

  - To support the democratisation of local spaces by developing and strengthening the capacities for design proposal into popular organisations to put forward and negotiate.
  - To support the empowerment of popular organisations within local governments, providing them with the necessary technical support to ensure governability and sustainable development at the local level.

Preliminary conclusions for a proposal of co-operation for development

Supporting development by strengthening social actors and enabling them to influence the organisation, distribution and articulation of social micro-scenarios is not an easy task. As was mentioned earlier, unless the counterparts address it responsibly, this task may turn into a fallacious proposal that instead of contributing to democracy and local develop-ment, it might abort organisational processes that would be more influential from the periphery of power institutions.

At the local level, articulating and co-ordinating the positions of dif-ferent actors, who often have conflicting interests, requires having clear proposals. To be able to talk about local development, one must take into account an array of aspects ranging from physical territorial determinants to the local space. This is (the place) where the various actors come together, fall into order and articulate themselves in physical, social and economic terms, within a regulatory institutional framework established by the system. The co-ordination and organisation of the actions of the different agents committed to a common development project depends, to a great extent, on the effectiveness of local administrators to develop and negotiate proposals, as well as to create a platform of human, financial and technical resources to balance immediate and long-term needs. Local development cannot entertain the notion of planning any undertakings for tomorrow, because the frustrations of the most destitute sectors demand immediate answers. Allies must be sought out, both in the local context and beyond, to conduct fund-raising strategies and to overcome the technical deficiencies that, at the beginning of the process, are charac-teristic of most local administrations.

At this point, important issues emerge that must be dealt with in the context of the process of empowerment of social organisations. Some of them are more important and relevant than others (depending on the characteristics of the various groups and their local and national contexts), and international co-operation must incorporate them into any initiatives designed to support the popular organisations. Such issues include, among others:

- *Design of local development strategies*: To conceive and design on a par-ticipatory basis a new territorial arrangement, including its productive potential, in the context of a distributive concept, friendly technologies and sustainable proposals.
- *Management and control of natural resources*: To have an inventory of all existing and potential resources in order to regulate and control their exploitation.
- *Sustainable development and ecological balance*: To ensure that economic activities will follow rational resource management procedures. This is a particularly important subject, because the local leader operates in a regulatory framework established by national power institutions; there is, though, local regulatory machinery that he/she must take into ac-count. Otherwise, the local leader should contemplate negotiating with any agents responsible for polluting activities, with a view to enforcing regulations on a gradual basis.
- *Cultural rights*: Full respect for ethnic minorities and social groups and enhancement of a local identity and collective self-esteem as a means of bringing people together and fostering communal attitudes.

- *Labour and environmental legal regulatory frameworks*:   Knowledge of and advice on legal regulatory frameworks is required. The legal issue is key to establishing what possibilities are available for the administration to engage in regulatory activities and at what level they are able to influence.
- *The right to educational and health services*:   Besides existing public programmes, bringing in co-operation projects on preventive health and education, especially those that target children and young people. To enhance the technical capabilities of the latter. To establish an alliance with local private enterprises to provide practical training to prevent qualified human resources from migrating.
- *The right to basic goods: air, water, energy (electricity, gas, etc.)*:   Unpolluted air and water are increasingly scarce. These are relevant issues, to be kept in mind by those who gain access to local power, since public services generally have limitations.
- *Local-level gender policies*: To promote women's participation in local power spaces, as well as meetings and other activities, refraining from excluding men.
- *Consumer rights*:   To support policies to safeguard consumer rights in the face of a marketplace that attempts to monopolise both the attention and resources of local families.

In the age of globalisation, local administrations and real popular empowerment require the ability to collect and manage a great deal of information, as well enabling the popular organisations to use modern communication tools. They also need to strengthen their capacities to develop and negotiate proposals. And all of the above should be achieved without neglecting internal democracy and organisational strengthening.

From the point of view of the co-operation, the goal of improving the living standards of the popular sectors, depending on their degree of empowerment, requires the complementing of the existing supports in the area of organisational strengthening. This embraces initiatives aimed at strengthening autonomous strategic capacities in the popular organisations with a view to enabling them to participate in the democratic exercise and influence the development of their own local spaces.

Local power is a strategy in process

In most cases of popular empowerment, and especially in Ecuador, where the victorious candidates have already been sworn into office, there is great concern with respect to demands in terms of information management, technical know-how and capacities to develop proposals. No such concern was expressed in Chile up to 1996, but in 1997, officials in local administrations asked for support with respect to strategic development in the environmental area, and a year later they requested assistance to conduct research to examine the poverty of local conditions thoroughly.

Supporting the exercise of local power implies institutional adjustments

In this respect, we should add, and at the same time differentiate, the meaning of recognising the institutional nature of local governments as part of the bureaucratic system, including their weak points, technical deficiencies and poor resources. This question has been widely raised and discussed in the case of Coronel, where NPA co-operation has contemplated, at different stages, supporting proposals asking for institutional strengthening and strategic design, while refraining from accepting the system as such (i.e. resorting to external support and consulting jobs). In Ecuador, we are currently supporting technical strengthening activities in three municipalities.

Maintaining social organisations as a subject of support

The institutional set-up, represented by municipalities, district councils and the like, is a tool to exercise local power on behalf of the organisations. In this sense, any methodologies to make institutional adjustments must be designed directly by the social organisations, and any support should be restricted to this context.

Reflection and exchange of experiences

The dynamics of the majority of the experiences of popular organisations and local administrations are determined by the urgency and precariousness of the various contexts. A contribution from international co-operation may include fostering the creation of opportunities for reflection on a regional level, to engage in a retrospective critical exercise about their proposals, as well as to learn from the different alternatives generated by the organisations in the various national and subregional contexts, to face the assault of globalisation on the popular world.

Annex: Reflections and results of the seminar. Social organisations and municipal governments held in Coronel in January 1996

Presentation

I believe that it is worthwhile to include this brief document, because I compiled it in an effort to remain faithful to the participants' expectations and reflections. Although more than two years have gone by since then, many statements in this document continue to be valid. Besides, it was after this seminar that we observed the levels of empowerment that the organisations can achieve and understood that the option of gaining power by means of municipal or regional administrative institutions is not always the right way. Depending on the type and characteristics of the social organisation in question, the degree of influence may decrease as a result of its positioning in the locus of power. While it may be too circumscribed, this document also bears witness to the joint learning process that NPA and its local counterparts have gone through.

Some important remarks

Participants from the 9th Region (Temuco; fishermen and indigenous people)

In general terms, we observed three levels of expectation among participants from the 9th Region, different according their respective positions and role on the issue of municipal elections as a proposal for the organisations.

Fishermen's leaders wanted to learn about the Coronel experience as a development alternative in an area of extreme poverty, trying to understand the similarities that might be adapted to their own needs as well as to their own conditions of poverty, but also in terms of the role that local fishermen might have played in Coronel's concerted action for development. They were able to realise the complexity of the challenge and the limits of embracing an electoral option as a sectoral trade union, taking into account that there are several sectoral and territorial organisations as players pushing for special attention within the scope of local development. They understood that, in spite of small-scale coastal fishermen existing in Coronel, they played a limited role in the efforts to redefine the productive perspective of the municipality, and that the social movement was led there mainly by coal miners (actually former miners) in its most critical stages.

Consultants, from a more technical perspective, requested to join the discussion of the way to apply and gain access to other public funds for municipalities. They had some difficulties understanding that negotiations over funds from municipal governments and from social organisations are conducted on a totally different basis. In spite of the problems involved in this issue, the municipal governments are in a better position as long as they manage to combine a good technical administration with a proper strategy in terms of lobbying and political bargaining. Consultants resented a failed attempt to negotiate packages of projects with the regional government. As a general comment, we may note that they initially failed to maintain some distance from the confrontational strategies of small-scale fishermen, often twisting the substance of the discussion of the main topics of the meeting. They also made a fairly extemporaneous request for a discussion on planning.

Having internalised their most relevant needs in terms of the electoral situation and the local context, the *candidates* showed openness and were eager to learn from the local experience. Eventually, they were shocked by the complexity of the challenge, acknowledging that they were not properly aware of it, and appreciated receiving documents and input from the discussion concerning their real chances in the coming elections. In informal asides, they got in touch with people in Coronel and arranged for some future informal advisory support.

Participants from Coronel

Municipal leaders (the mayor and his *chef de cabinet*) showed openness and frankness in discussing their successes and failures without prejudice. They underlined the point that no experience can be replicated in its entirety, considering that each territory has its own particularities, adding that their contribution stopped at telling their story candidly (both the official and the unofficial one), leaving the discretion of the audience as to whether there were lessons to be learned from it.

On the other hand, the general discussion produced some material regarding the future development of Coronel in terms of growth-generated impacts, especially in the environmental area. Some proposals were made to be included in the mayor's agenda for a possible second period of administration, for which he was standing at that time.

CEPDEPO's director discussed the relationship of the project with the social movement into the commune, before and after entering into the local government. The intervention of the city's planning secretary offered a complete and illuminating presentation – supported with visual materials – on municipal funds open to public bidding.

Participants from Santiago

Representatives of CENFOCAR and Raíces, two counterpart groups invited from Santiago, who participated basically as *observers*, played fairly passive roles. However, they blended into the group and exchanged views with other delegates in informal asides. A representative of CEN-FOCAR established some links with an indigenous leader with a view to future meetings with the organisation in Santiago (in a poor sector where they work). As a *technical observer*, Raíces provided important input to play against the conclusions of this report and stated that the meeting widened its views on some aspects that had been taken into account in some decisions made by Raíces.

The leaders of the Toma, an organisation for the homeless in Santiago, played an active and critical role, balancing the meeting during the debate with the fishermen's advisors. They came to the meeting eager to learn essentially from the Coronel story, which they praised. They seized the occasion to enter into formal agreements concerning exchanges and advisory support with leaders in Coronel.

Main conclusions

Issues and challenges for social organisations

The main objective of gaining access to local governments is important for social organisations on the basis of taking control over decision making and investment spaces in order to achieve improved standards of living for the poor sectors. The popular movement should not confuse its fight against the system and the present development model with the fight to achieve changes in their living conditions in the short and medium term. This was expressed by José Luis Flores.

> For the popular organisations, the spaces could be different, but the deep problem is always the same. In the 9th Region the need is to take control over marine resources, which entails winning guarantees for the fishermen. The preservation of Mapuche territories and culture

requires the recognition of the rights of the indigenous people. The town of Coronel died due to the coal crisis and the lack of resources. Santiago's La Pintana is a poverty ghetto. Homes were built there but other needs of the people were disregarded, such as the creation of productive activities to provide jobs. The story of La Toma confirms that it is important for social movements to gain a place in local governments. But they must make sure that their people will not be co-opted by the central government and that they will be able to fight for improvements.

(From an intervention by José Luis Flores, the president of the organisation of the homeless, 'Esperanza Andina de Panalolén')

- The fishermen's organisation must evaluate the different ways of gaining access to municipal government. The office of mayor, the City Council or CESCO (the Economic and Social Committee, consisting of social leaders in each municipality) are the gates that social organisations should be able to cross in order to share in local power administrations in Chile. Each one of them has its pros and cons. Each organisation and its candidates must assess issues such as electoral coalitions, technical support and capabilities. They should also determine what the appropriate level is for them to engage in local power. If their fight involves a particular group or sector, they may have a better chance of exercising influence through a CESCO. They should also take into account that popular demands at the municipal level are not restricted to a single productive group.

The municipal issue is very complex. There are advantages and many disadvantages. Serious doubts remain. ... What role will the Federation of fishermen role regarding Saturnino's electoral bid?

(From an intervention by Yerko Castro, technical consultant to the 9th region – Temuco – Fishermen's Federation)

- Problems affecting local governments are caused by legal issues as well as the lack of support from both Congress and the national government.

... legal hurdles, lack of time. We are always summarising; even our family life becomes telegraphic. People are exhausted, there are no allies in Congress, we still have to persuade Government officials. Municipal governments have little decision-making power; national officials are mostly concerned with cutting heads.

(From an intervention by Eduardo Araya, chief of cabinet and political assessor of the mayor of Coronel – later on he became the regional councillor for Bio Bio region in southern Chile)

- The transition from the organisation to the municipal government requires a participatory management as well as strengthening the social movement and being able to incorporate it into the administration.

... it was tough. We were a group of crazy guys coming down from the Cerro de la Virgen to start a fight. ... We got into the Party. ... Now the approach is to address all the problems in public, facing the people, letting them know whatever we are doing. Since people have no money to buy papers and we have no money to print anything, we took and wrote on the walls, stop at the corners to chat with the people in the street, get on the local radio.

(From an intervention by Eduardo Araya)

- The pros and cons of the candidates were assessed in general terms.

Saturnino Ulloa, a veteran leader of miners and fishermen (who failed in the 1996 elections), is influential in his municipality and holds in his hands the votes of fishermen and the people in his home area (La barra del Toltén). Yet these votes do not guarantee him a seat, due to electoral law restrictions. Therefore he must choose a political party. He must get wide-ranging technical support in order to cover the complexity of his area's demands. He has many strengths as a candidate; he appears to have a greater chance to become a council member (thanks to the number of votes he controls). Support provided to him by advisors with the fishermen's federation must be revised. They must define their role and the kind of support that they can provide, refraining from seeing fishermen as those who run the local government. The leader's social influence is due to the fact that his community role goes beyond the fishing sector. He must draft his agenda for development taking that into account.

Lorenzo Aillapán is an indigenous leader in Puerto Saavedra, who also failed in the 1996 elections. He has developed a strong background in the area of preserving and promoting Mapuche culture and can count on vast local, national and international contacts. He may be able to put together a team with other Mapuche Indians and trustworthy white advisors; he holds in his hands the votes of the Mapuche community (over 40 per cent of the electorate). He could make a bid as an independent candidate, although this would be a risky move due to the flaws in the electoral system. He has been advised to chose a party in order to get elected. He may be able to become mayor. His agenda for the Mapuche people should take into account both their problems and their limits.

Industry and forestry operations are belated conquest for the Mapuche people [referring to the negative cultural impact of Spanish conquest]. Our people are intimately linked to the land. Presently the land problem is a humiliating issue for us. ... The problem is that there are three kinds of Mapuche: those who stayed on their land, the intellectuals and those who reject their land and have changed their names.

(From an intervention by Lorenzo Aillapán)

- Political campaigns demand that the candidate run as a winner from the very beginning, showing a high level of confidence and self-esteem, as well as the ability to gain the support of different social groups as allies:

The candidate must feel that he/she will win, that he/she will get the first relative majority. Whoever is perceived as a bad runner up candidate turns into an instant loser. ... He/she must attract people, must be convinced of being a winner, feel like a winner and appear like a winner. ... Women voters are a majority, they represent more than half of the electorate. They must be included. ... When they make a decision they hardly go back on it.

> (From an intervention by René Carvajal, rejected in
> the local polls in Santiago)

In a politial campaign, the main meaning is to add people not to marginalise them (Clemira Pacheco, executive director of CEPDEPO).

Issues and challenges in co-operation for development

Co-operation organisations must preserve the flexibility of their support strategies, but they must also supplement their follow-up work with efforts aimed at detecting any weaknesses and using the social resources that they can muster through their counterparts, with respect to their decisions and needs. This is neither a simple barter deal nor an imposition on the part of co-operation, but rather an imaginative effort to channel acquired knowledge and to reinforce certain processes in a timely fashion. This seminar allowed us to determine that the CEPDEPO-Municipality into local power and development experience is in itself a pedagogical resource/tool, as much as the Toma (the organisation of Santiago's homeless) may become – all these without losing sight of the role that international co-operation must preserve as an external facilitator.

The issues of local development and municipal elections have turned into recurrent subjects for the social organisations, demanding international co-operation. In the case of NPA Chile, this is an effort to follow up the electoral processes as a preliminary diagnostic about the roots of organisations and local power, in other words, to detect in which other national contexts there have been proccesses that may offer lessons capable of widening our understanding of these issues.

The jury is still out on whether to consider having discussions with differentiated actors, considering that the concerns and expectations of the social organisations and their leaders do not always coincide with those of their internal or external advisors. This problem may require the adoption of a different kind of methodology.

Discussion opportunities necessarily should include the encouragement of affections, to the extent that we work with people who live under adverse conditions. Social learning processes need the knowledge and comunication space, but also need to become part of our emotional experience in order to be internalised positively. In this fashion, social organisations may start to acknowledge each other.

Participants

Coronel

René Carvajal   Mayor of Coronel, former president of the Centre of Popular Promotion for Coal Zone (CEPDEPO).

Eduardo Araya   Chief of cabinet of municipality of Coronel, political assessor, and regional councillor for 8th Region, Bio Bio.

Clemira Pacheco   Executive director of CEPDEPO.

Fishermen Federation 9th Region; Araucania, Temuco

Saturnino Ullo   Former president of federation. At that time was a candidate for mayor of his municipality in Toltén.
Pedro Morales   Leader of federation.
Lorenzo Aillapán   Indigenous leader, 9th Region. At that time was a candidate for an Indian municipality in the 9th Region.
Luis Báez   President of federation at that time.
Pablo Porras   Councillor of federation.
Yerko Castro   Councillor of federation.

Organisations from Santiago

José Luis Flores   President of homeless organisation Campamento Esperanza Andina (Toma de Peñalolén).
Gonzalo Concheso   Leader of Campamento Esperanza Andina.
Denisse Araya   Executive director of the NGO Colectivo Raíces from Santiago.
Roberto Alvarez   Youth leader of Centre of Popular Training Aracely Romo (CENFOCAR).

Participants from NPA

Nelson Soucy.
María Verónica Bastías

References

Alburqueque, F. (1995) *Espacio, Territorio y Desarrollo Económico Local.* ILPES, Santiago de Chile.

Alburqueque, F. (1997) *El proceso de Construcción Social del Territorio para el Desarrollo Económico Local.* ILPES, Santiago de Chile.

CASEN (1998) *Caracterization socioeconomics national.* Division de Planificacion, Santiago de Chile.

CEPAL (1997) *La Brecha de la Equidad. América Latina, el Caribe y la Cumbre Social.* Comisión Económica para América Latina y el Caribe, Santiago de Chile.

CEPAL (1998) *Panorama Social de América Latina 1997.* Comisión Económica para América Latina y el Caribe, Santiago de Chile.

CEPAL/IIDH (1997) *La Igualdad de los Modernos. Reflexiones acerca de la Realización de los Derechos Económicos, Sociales y Culturales en América Latina.* Comisión Económica para América Latina y el Caribe/Instituto Interamericano de Derechos Humanos, Santiago de Chile.

Coraggio, J.L. (1994) Del sector informal a la economía popular: un paso estratégico para el planteamiento de alternativas de desarrollo social, *Revista La Piragua* **8**, CEAAL, Santiago de Chile.

Freire, P. (1972) *Pedagogy of the Oppressed.* Herder & Herder, New York.

Galilea, S. (1993) *Los Procesos de Descentralización en América Latina: Análisis y Perspectivas.* Santiago de Chile.

Hagmann, K. (1996) *La Sociedad Civil en la Cooperación entre Europa y América Latina: Las Organizaciones No Gubernamentales.* Instituto de Estudios Estratégicos e Internacionales de Lisboa, Ponencia. Prepared for Segundos Encuentros Euro-Latinoamericanos, Río de Janeiro, December 1996.

IMEP (1997) *Estrategias e Instrumentos para el Alivio de la Pobreza Extrema.* Instituto Mexicano de Estudios Políticos A.C., México.

McIlwaine, C.J. (1998) Contesting civil society: reflections from El Salvador, *Third World Quarterly* **19**(4): in press.

NPA (1993) *Solidarity Across Borders. Policy for International Activity in the 1990's.* Norwegian People's Aid, Oslo.

NPA (1994) *Assistance to Self-reliance. Action Plan 1995–1999*, Norwegian People´s Aid, Oslo.

NPA (1995) *Preliminary Strategic Document for NPA South America.* Norwegian People´s Aid.

PNUD (1998) *Desarrollo Humano en Chile – 1998. Las paradojas de la Modernizació.* Programa de Naciones Unidas para el Desarrollo, Santiago de Chile.

Simon, D. (1998) Rethinking (post)modernism, postcolonialism and post-traditionalism: South–North perspectives, *Environment and Planning D: Society and Space* **16**(2): 219–245.

Valarezo, G.R. (1996) *Movimientos Sociales y Gobiernos Locales.* Prepared for La Reunión de Diputados del MUPP-NP auspiciado por la Fundación Peralta-ILDIS, Ecuador, July 1996.

Conclusions

CHAPTER 12

Conclusions and prospects

David Simon and Anders Närman

We begin our conclusion by referring to two television items, a clear indication of the power of the visual image in reflecting and shaping popular perceptions, in this case of issues surrounding development aid. These examples represent two contrasting visions, which serve as metaphors for the central concerns and arguments of this book.

On Swedish television in the early 1990s, the then minister for development assistance, Alf Svensson, was bitterly attacked over the country's aid policies. In defence, he claimed that aid could in part serve as a way of repaying the US$100 billion that we in the North gain from trade with the South! This approach is sure to perpetuate global inequalities, so it raises the question of whether we should not examine problems in current trade structures and patterns more closely if we are serious about promoting greater equity. Later in the TV debate, Svensson wanted to illustrate what he regarded as a good aid project and produced a bottle of worms removed from boreholes in India. This, he argued, was a practical example of what aid was really about. So, we are supposed to clear Indian boreholes! Have we come no further in the last fifty years? What evidence is there of the much-hyped objective of giving help to promote self-reliance?

British television news at the beginning of August 1998 featured the Secretary of State for International Development, Clare Short, visiting development projects supported by the British government in postwar Mozambique. One particular project highlighted involved the provision of a single goat to households which had lost all their livestock through displacement. The goats were donated without charge on condition that the recipients would care for them and repay the project with one goat kid when the animals produced offspring. As with similar successful projects elsewhere, the objective is to facilitate restocking and thereby to promote self-reliance.

Part of the project's appeal is its simplicity. However, it also embraces at least three other important principles. First, it involves the recipients as active participants in the project; they play the central role and have the responsibility of rearing the animals and ensuring that they reproduce. This complies with current approaches to good practice, which seek to avoid generating or perpetuating dependence and passivity. Second, it embodies the principle that the recipients of development assistance have an obligation to repay a proportion of their yield that is roughly equivalent to what they initially received, and when they are in a position to make such repayment without compromising their gains. Third, the act of repayment is not to please the donor but is the means by which other destitute households are able to benefit. As such, it resembles a rotating credit scheme, where the start-up capital is provided by a donor, except that the capital in this case is livestock not money. It is also a locally appropriate variant of democratic accountability.

Hence this seems a model development project in the late 1990s. No doubt that was why it received prime-time TV coverage, conveying the message that British taxpayers' money in the form of development assistance is being well spent in some of the poorest countries of the world. However, one of the most striking features of this project – and something not mentioned in the news item – is that it is far from new. In fact, it is arguably the oldest recorded type of aid, illustrating superbly well the biblical parable that by giving a person a fish you feed him/her for a day, but by teaching him/her to fish you feed him/her for life.

This allegory implies that, although fashions – in development theory and practice as elsewhere – come and go, the simplest and most important often prove resilient, enduring without interruption or slipping out of the collective consciousness to await subsequent resurrection. As we have documented in this book, the history of development over the last fifty-odd years has been more complex than some critics would acknowledge. New ideas and grand schemes have been attempted, out of naivete, a genuine desire to 'do good' or geopolitical strategy. Mainstream official policies have undergone a series of modifications in an effort to overcome past shortcomings and to meet changing perceptions or priorities. Many of the objectives or concerns taken up by the major agencies were initially articulated by radical fringe groups but then struck a chord with informed opinion growing frustrated with the apparent failure of existing practice. This raised the stakes, precipitating efforts to assimilate aspects of these new agendas by the mainstream so as to project a progressive and responsive image. Ironically and probably inevitably, however, the selective adoption of radical alternatives has seen their emasculation, rendering them increasingly less radical and more accommodationist. This adaptive strategy generally draws the sting of opposition, except from the margins, where radical critics highlight the compromises of principle involved in such mainstream assimilation.

Nevertheless, beneath such adaptive change, with all the associated publicity, the basic tenets of neoclassical economics and modernisation theory have endured remarkably intact. Indeed, they re-emerged with renewed vigour in a frighteningly simplistic (some might say 'pure') form in the early 1980s, leading the neoliberal challenge to Keynesian and neo-Marxist formulations, which embody a substantial developmental role for the state and which had held sway through the late 1960s and 1970s. This challenge claimed that the main problems facing countries of the South were attributable to state inefficiency and profligacy, and high levels of corruption. In order to revitalise these economies and promote renewed economic growth, it was necessary to roll back the state and 'get prices right'. This is the stuff of which structural adjustment programmes and their offshoots were made. The pendulum is beginning to swing back towards softer, more socially sensitive and nuanced approaches, but there is little sign that the fundamental ideology of development embodied by the multilateral agencies and some major donors is changing.

In various complementary ways, the chapters in this book have conveyed some of the problems and limitations inherent in strict modernisation theory and its applications, none more so than the SAPs. However, it is worth pondering just why this theoretical formulation should prove so endurable in the face of such sustained critique and mounting evidence of its limitations and of widespread development failures. The answer is complex but can probably be reduced to four key points. First, it remains consistent with the dominant neoclassical economic ideology of development within the USA and most of Western Europe, not least under social democratic governments in the latter. Second – and this relates to the biblical parallel cited above – it is a very simple and universalistic formulation. Third, there have always been sufficient apparent successes to point to as sources of vindication. Probably the most powerful of these is the demonstration effect of higher living standards and the material trappings of modernisation, which David Simon discussed in Chapter 2. Once absorbed by the indigenous elites during late colonialism and given full play in government policies of virtually all ideological persuasions following independence, the logic has become ingrained. Reinforced by advertising and rising living standards among some social groups and in some countries, it is undeniable that the aspirations of most poor people are framed at least in part by the conspicuous consumption of their wealthier compatriots. Frustration at prolonged inability to meet their aspirations, particularly when recession or structural adjustment policies and their associated hardships bite deep, often give rise to protests and riots. In early 1998, the venting of such anger was sufficient to bring about the political demise of President Suharto, the longtime autocratic ruler of Indonesia.

The fourth factor in the survival of the underlying modernisation ethos is the astonishing speed and inventiveness (some would say desperation) with which the Bretton Woods institutions rally to its defence, even in the face of overwhelming odds. Hence, as Chris Dixon has demonstrated so effectively in Chapter 9, when previously vaunted success stories go sour, the very policies hitherto held up to other countries as the models to emulate are reinscribed as inappropriate. So, whereas we were told for so long that the Asian Tigers represented what can be achieved with market-led, export-oriented industrialisation, we are now blithely being asked to believe that the crises currently being experienced in these countries reflect inappropriately high levels of state interference in their markets. Critics of the conventional wisdom have long been highlighting the very important roles of the state – both directly and indirectly – in launching and maintaining aggressive economic growth. However, there must be a credibility gap when IMF and WB reports instantly convert yesterday's wisdom into today's problem, without so much as a hint of self-castigation or culpability for the previous errors of ideology, policy or judgement. Once again – as with the debt crisis, for which the new medicine became SAPs – the crises are portrayed as being entirely the fault of the countries concerned. Yet nowhere do we hear guffaws of derision or incredulity. These reports somehow achieve a certain status and credibility, and thus the (surely discredited) ideology being defended so blatantly seems to be surviving yet again.

One of the principal themes addressed in this book, especially by Anders Närman (Chapter 7), and Ali de Jong *et al.* (Chapter 8) but also to some extent by María Bastías (Chapter 11), is the nature of evolving debates regarding development aid and the policies formulated to give effect to changing priorities and objectives. These contributors have indicated how relatively small donors like Sweden and the Netherlands have come to follow reasonably closely the trends set by major multilateral agencies and USAID since the debt crisis. There is today far less differentiation in donor policies than during the 1960s and 1970s, when a small group of 'progressive' social democratic states pursued substantially different agendas and were at times openly critical of the view from Washington and New York. For example, Anders Närman argued in Chapter 7 that Sweden selected its main co-operation partners partially to express disapproval of US foreign policies. In many respects, the narrower range of perspectives probably facilitates multilateral programmes like the Healthy Cities initiative discussed by Ton van Naerssen and Françoise Barten (Chapter 10). However, it is increasingly in the NGO sector that important mould-breaking initiatives can (and perhaps have to) be pursued. This sector is extremely diverse and care is needed not to lose sight of this, since there is no inherent reason why all NGOs or CBOs should necessarily be efficient, effective or progressive. This assumption has all too readily been made by donor governments keen to promote civil society as part of their wider agendas. One particular category of initiative is co-operation for development through people-to-people organisations like Norwegian People's Aid, where donor funding and advice are articulated through local CBOs run by appropriately skilled nationals of the countries concerned. María Bastías has provided us with a frank first-hand account in Chapter 11.

At present, some donors are again rethinking their policies. One of the more substantive recent policy changes has occurred in the UK, following the Labour Party's dramatic election victory in May 1997. The status of the Overseas Development Administration (ODA) was immediately upgraded to that of a full government department, the Department for International Development (DFID), with the outspoken Clare Short as secretary of state. Her stance to date is reminiscent of the influence on aid policies of Jan Pronk and Olof Palme (see Chapter 7). Statements about the need to refocus aid policy were followed by the publication in November 1997 of the White Paper on International Development (DFID 1997), which sets poverty alleviation as the central objective. Existing commitments are being re-evaluated, and all new research and funding proposals are to be appraised in that light. While it is far too early to form a meaningful judgement, not least in regard to how substantial a change to *practice* this actually represents, the new commitments have generally been positively received, and the UK has certainly been able to voice its concerns about development issues from a more credible position and to take the lead on issues ranging from debt write-off negotiations within the G-7 group of the world's wealthiest countries to the banning of the manufacture, stockpiling and use of land mines.

The contributors to Section 1 of the book have advanced important and provocative arguments about the nature of development studies and its practice, both in and of itself but also in relation to academe in general. Development theory is in ferment, led by exciting debates about the relevance of postmodernism, postcolonialism and other perspectives to countries and societies of the South. In this endeavour, development studies is engaging firmly with other social and interdisciplinary sciences (see Simon, Chapter 2), and is well placed to take a lead by virtue of its unique position of being able to draw upon first-hand experience in cross-cultural situations mediated by vast inequalities of economic, social and political power. Yet, despite all the political correctness and apparent concern for cultural sensitivity and 'distant others' among academics in various disciplines, development studies and area studies – and their practitioners – are still regarded as marginal to the mainstream discipline-based endeavours of the academic world in the North. Mike Parnwell (Chapter 4) has made an eloquent and personalised case for this to change, something that would benefit all concerned. As Rana P.B. Singh has argued in Chapter 3, one critical way of promoting substantial change in both academic and practical terms would be to re(instate) an explicit spiritual dimension to our lives and to development efforts on the ground. Although India was his example, the message has universal relevance. Reg Cline-Cole's analysis of the academic practices of geographers in North America and the UK in relation to Africa and Africanists complements that of Mike Parnwell, indicating the vastly different but interlinked world inhabited by geographer colleagues in Africa. Genuinely felt concerns to help often lead to paternalistic or inappropriate actions. In Chapter 6, Dan Tevera gave a very personal account of how these vast differences in resourcing and context affect the individuals from the South who undertake (mainly) postgraduate studies in the North and then return to their home countries to deploy their newly found skills for the benefit of their countries and compatriots. Awareness of some of these issues and problems is hopefully growing among those with the resources and linkages who are able to make a difference in future.

So what is the future likely to hold? We have deliberately avoided the temptation to indulge in crystal gazing in the name of some profound fiftieth anniversary, end-of-century or millennial insight. All of us who have contributed to this volume would probably agree that it would be foolhardy to predict some great leap forward or a sudden 'revelation' leading to the narrowing of North–South gaps and a lessening of the contradictions of development. The current world system may be 'post-Cold War' but nevertheless remains riven by conflicts – some indeed having been suppressed during the Cold War – and dynamic in every sense. Information technology and the processes of globalisation which it is accelerating mean that news is instant, so that turmoil in one country or region is immediately reflected in stock market jitters around the world, for example. The fortunes of different countries and regions may also fluctuate increasingly rapidly, as the 1997/98 crises in Southeast Asia have demonstrated. Simultaneously, however, the World Bank has begun to talk more positively about Africa's prospects, buoyed by apparently more rapid rates of economic growth as measured by GNP and GNP per capita. This is one factor that has led to recent discussion of a no-aid future, an objective that would in itself mean an immediate but temporary increase in aid.

We would urge caution, both because of the well-known limitations of such indicators but also because long-entrenched politico-economic structures of dominance and subordination are difficult to reshape more than marginally. Finally, as we write this in August 1998, Angola, one of the countries about which the World Bank has recently enthused, is lurching rapidly back to bloody civil war just as meaningful peace seemed within grasp. The searing famine in Bahr-el-Ghazal province in southern Sudan is very largely the result of that country's vicious civil war. In addition, reports of increased fighting in the Democratic Republic of Congo, Rwanda, Burundi and even Uganda (the perceived development success story of the 1990s), as well as the recent outbreak of hostilities between Ethiopia and Eritrea, point to ongoing instability with unknown consequences. The same is true in Cambodia, Afghanistan, Kosovo/Yugoslavia and several other countries. Unlike traditional conventional wars, which pit armed forces against one another, a hallmark of current armed conflicts is the conduct of irregular warfare either targeting civilians deliberately or being conducted in the midst of civilian concentrations so that they become the principal victims. The distinction between domestic and cross-border conflicts is also now increasingly blurred. Development is impossible under such circumstances; survival and emergency assistance inevitably take priority.

Nevertheless, we end on a more optimistic note. Development studies (including development geography), so long something of a cinderella in academic terms, has clearly now come of age. Those who confidently predicted its demise a few years ago as a result of the postdevelopment or antidevelopment critiques were mistaken. Certainly there was and remains a need for change, and continual refocusing in order to transcend previous limitations of the sort discussed at length by many contributors to this book. However, the basic importance of and need for appropriate understanding of the issues facing the majority of the world's population has not changed and will not go away, even if some Northerners choose to opt out in the name of cultural relativism. Hence development studies remains a very popular student choice; the demand for graduates with these skills and expertise remains buoyant; and research, publication and theoretical debate are flourishing. The subject area has proved equal to the challenge, and is emerging reinvigorated, even leading the way in terms of more realistic poststructuralist theoretical formulations, which really have the potential to make a difference to both policy and practice.

Reference

Department for International Development (1997) *Eliminating World Poverty: A Challenge for the 21st Century* (White Paper on International Development). DFID, London.

Index